One-Cocycles
and
Knot Invariants

SERIES ON KNOTS AND EVERYTHING

ISSN: 0219-9769

Editor-in-charge: Louis H. Kauffman *(Univ. of Illinois, Chicago)*

The Series on Knots and Everything: is a book series polarized around the theory of knots. Volume 1 in the series is Louis H Kauffman's Knots and Physics.

One purpose of this series is to continue the exploration of many of the themes indicated in Volume 1. These themes reach out beyond knot theory into physics, mathematics, logic, linguistics, philosophy, biology and practical experience. All of these outreaches have relations with knot theory when knot theory is regarded as a pivot or meeting place for apparently separate ideas. Knots act as such a pivotal place. We do not fully understand why this is so. The series represents stages in the exploration of this nexus.

Details of the titles in this series to date give a picture of the enterprise.

More information on this series can also be found at http://www.worldscientific.com/series/skae

K&E Series on Knots and Everything — Vol. 73

One-Cocycles
and
Knot Invariants

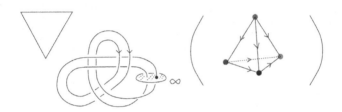

Thomas Fiedler
Université Paul Sabatier, France

World Scientific

NEW JERSEY · LONDON · SINGAPORE · BEIJING · SHANGHAI · HONG KONG · TAIPEI · CHENNAI · TOKYO

Published by

World Scientific Publishing Co. Pte. Ltd.

5 Toh Tuck Link, Singapore 596224

USA office: 27 Warren Street, Suite 401-402, Hackensack, NJ 07601

UK office: 57 Shelton Street, Covent Garden, London WC2H 9HE

Library of Congress Cataloging-in-Publication Data

Names: Fiedler, Thomas, author.

Title: One-cocycles and knot invariants / Thomas Fiedler, Université Paul Sabatier, France.

Description: New Jersey : World Scientific, [2023] | Series: Series on knots and everything, 0219-9769 ; vol. 73 | Includes bibliographical references and index.

Identifiers: LCCN 2022026784 | ISBN 9789811262999 (hardcover) | ISBN 9789811263002 (ebook for institutions) | ISBN 9789811263019 (ebook for individuals)

Subjects: LCSH: Knot theory. | Invariants. | Combinatorial analysis.

Classification: LCC QA612.2 .F546 2023 | DDC 514/.2242--dc23/eng20221007

LC record available at https://lccn.loc.gov/2022026784

British Library Cataloguing-in-Publication Data

A catalogue record for this book is available from the British Library.

For any available supplementary material, please visit
https://www.worldscientific.com/worldscibooks/10.1142/13044#t=suppl

Desk Editors: Nimal Koliyat/Adam Binnie/Shi Ying Koe

Typeset by Stallion Press
Email: enquiries@stallionpress.com

Printed in Singapore

For Séverine with all my love

Was lange währt ist endlich gut
geworden
(still old-fashioned saying)

Preface

This book is about classical knots, i.e., smooth oriented knots in 3-space. We introduce discrete combinatorial analysis in knot theory and use this new technique in order to construct combinatorial 1-cocycles. They can distinguish sometimes knots up to isotopy, as well as loops of knots in moduli spaces up to homology.

Knots are usually given by diagrams and we could consider the moduli space M of all knot diagrams of classical knots. Many known knot invariants, e.g., quantum knot polynomials or finite type invariants, can be seen as *combinatorial 0-cocycles* on M. We associate a number or a polynomial with each generic knot diagram, and when two knot diagrams are in the same component of M, then these numbers or polynomials are the same. A generic isotopy of two knots corresponds to a generic arc in M. On such an arc, we see only a finite number of non-generic diagrams, which correspond to the well known Reidemeister moves. The most complicated one is the R III move, which corresponds to passing in the isotopy through a diagram which contains a triple crossing. A value associated with knot diagrams is invariant under this move if it satisfies the so-called triangle or Yang–Baxter equation. An important point is that all solutions of this equation are of a local nature: one associates with each of the three ordinary crossings involved in the R III move a matrix or an operator which does not depend on the remaining crossings of the diagram. This is where representation theory enters 3-dimensional knot theory. In this way, very many knot invariants were constructed. However, there are still not enough such knot invariants in order to distinguish all classical knots! Indeed, one of the most difficult problems is to distinguish the orientations of knots. It is known that all quantum knot polynomials fail to do this. Thirty years after their discovery, there are still not known any finite type invariants which could do that. Greg Kuperberg has shown that if finite type invariants fail to distinguish knot orientations, then they even fail to distinguish non-oriented knots as well. So, we are still very far away from the solution of the main problem in knot theory: *defining a system of calculable numerical invariants, which distinguishes all oriented smooth classical knots.*

A natural idea is now to use also higher dimensional cocycles on the moduli space M, in particular, 1-cocycles. However, Allen Hatcher has shown that if a knot is not a satellite, then its connected component in M has only a finite fundamental group. Consequently, each 1-cocycle on M, which takes its values in the integers or in integer polynomials, will represent the trivial cohomology class. Moreover, in a generic 2-parameter family of diagrams, we will see diagrams which contain a quadruple crossing. A 1-cocycle has to be 0 on the meridian in M of the diagram with the quadruple crossing (see the Introduction for the definition of the stratification of M). This means that the 1-cocycle has to satisfy the tetrahedron equation, which is a higher dimensional analogue of the triangle equation. The tetrahedron equation also has many local solutions coming from representation theory, but as explained above, all these solutions lead only to trivial 1-cocycles!

Consequently, we have to define a better moduli space than M and we have to construct global solutions of the tetrahedron equation, i.e., they should depend on all crossings of the diagrams and not only on the six ordinary crossings involved in the tetrahedron equation. It took us a long time to find out the right moduli space, called M_n, which will be defined in the following. A sub-set of its connected components is in 1–1 correspondence with knot types of classical knots. A 1-cocycle should now be a closed differential 1-form on M_n, which we integrate over smooth oriented loops in M_n. However, it seems to be extremely complicated to define such a 1-form on the infinite dimensional space M_n, where each point corresponds to a complicated knot diagram. In particular, the 1-form should take into account that the diagrams belong to M_n and not to M, otherwise it would again be trivial.

Luckily, combinatorics can sometimes replace not only algebra but analysis as well. We define *combinatorial 1-cocycles* on M_n. They can be seen as a discrete integration over loops in M_n. We associate nothing to ordinary knot diagrams in the loop but at each R III move in the loop it blows up to an integer contribution, and we sum up these contributions over the loop. Our strategy now is the following: for each generic point in M_n, there exist universally defined canonical loops such that if two points are in the same component of M_n, then the corresponding canonical loops for the two points are homologous. These canonical loops represent homology classes of infinite order for all knots different from the unknot. We now evaluate our combinatorial 1-cocycles on these canonical loops.

This allows us not only to distinguish homology classes of loops in M_n but also to distinguish sometimes connected components of M_n and hence knot types of classical knots.

In the book *Polynomial One-cocycles for Knots and Closed Braids*, we have laid the framework for constructing combinatorial 1-cocycles for classical knots (and for closed braids) and we have given first examples. This book is the sequel and we use our techniques to find more solutions of the global tetrahedron equation and to construct more combinatorial 1-cocycles for knots. In particular, we lift the whole Conway polynomial of a classical knot to a non-trivial combinatorial 1-cocycle in a certain moduli space of knots in the solid torus. The polynomial-valued 1-cocycle depends on two integer parameters $n > 1$ and $0 < a < n$. Remarkably, this 1-cocycle generalises straightforwardly the one from a lift of the coefficient of z^2 of the Conway polynomial by using *distinguished couples of persisting crossings*. We then construct another lift to a 1-cocycle of the coefficient of z^2 of the Conway polynomial by using *distinguished triples of persisting crossings* and we conjecture that this new 1-cocycle generalises straightforwardly to a lift of the whole Conway polynomial, exactly as in the case of the "distinguished couples of persisting crossings." We give this generalisation by an explicit formula, but we expect that the proof that it is indeed a 1-cocycle will be much more complicated.

There is a natural projection pr of the 3-space into the plane and knots in 3-space can be given by knot diagrams, i.e., a smoothly embedded oriented circle in 3-space together with its projection into the plane. It is well known that oriented knot types are in 1–1 correspondence with knot types of oriented long knots. We now close the long knot with the 1-braid to a knot in the standardly embedded solid torus $V \subset \mathbb{R}^3$, which can be seen as a long solid cylinder with the two discs at infinity glued together. Moreover, a (black board) framed long knot can be replaced by its parallel n-cable (with the same orientation of all strands) which we close to a knot in V by a cyclic permutation braid or more generally by an n-component string link T, which induces a cyclic permutation of its end points. The projection into the plane now becomes a projection pr into the annulus. We chose as generator of $H_1(V)$ the class which is represented by the closure of the oriented 1-braid. Hence, the knot in V, which is obtained from the n-cable, represents the homology class $n \in H_1(V)$; this is one of our two integer parameters. We consider the infinite dimensional space M_n of all knots K in V, which represent the homology class n such that the projection pr into the annulus is an immersion and such that there are no negative loops in $pr(K)$ (see the following). Moreover, we fix a point, called ∞, on the intersection of K with the disc at infinity and we add a new condition, called *separation condition*: a crossing cannot move over the point ∞ at the same moment as a Reidemeister III move happens somewhere in the diagram.

M_n *is called the moduli space of long regular knot diagrams without negative loops in the solid torus and which satisfy the separation condition.*

It is well known that the knot types in 3-space with fixed Whitney index (i.e., fixed rotation number of the corresponding oriented planar curve $pr(K)$) and with fixed black board framing correspond exactly to the connected components of long knots in 3-space up to regular isotopy. Consequently, they also correspond to the connected components of M_1. Given a generic knot diagram K in the solid torus V (i.e., an oriented embedded knot in V together with its projection pr into the annulus), we consider the oriented curve $pr(K)$ in the annulus. An *oriented loop in* $pr(K)$ is a piecewise smoothly oriented immersed circle in $pr(K)$ which respects the orientation of $pr(K)$. In other words, we go along $pr(K)$ following its orientation and at a double point we are allowed to switch perhaps to the other branch but still following the orientation of $pr(K)$. Naturally, a loop in $pr(K)$ is called *negative* (respectively, *positive*) if it represents a negative (respectively, positive) homology class in $H_1(V)$. One easily sees that $pr(K)$ contains only positive loops if and only if $K \subset V$ is isotopic to a closed braid with respect to the disc fibration of V and that knots which arise as oriented cables of long knots by the above construction (i.e., elements of M_n) never contain negative loops. To each crossing of a diagram in V, we can associated in a canonical way a homology class in $\mathbb{Z} \cong H_1(V; \mathbb{Z})$. We can fix such a homology class $0 < a < n$ as our second integer parameter.

The *peripheral system of a knot* is the triple which consists of the knot group (i.e., the fundamental group of the complement of the knot in S^3) and the homotopy classes in it of the meridian and of the longitude up to simultaneous conjugation. Waldhausen has proven that the peripheral system is a complete invariant for classical knots. The construction of our moduli space M_n uses the meridian because the solid torus V can be seen as the complement of the meridian in the 3-sphere, and the construction of our 1-cocycles uses homology classes in $H_1(V)$. It also uses the longitude because we have to replace our knot by twisted cables in the solid torus (in fact, our main 1-cocycles are not well defined for $n = 1$, hence using the longitude is essential too). The Conway polynomial is completely determined by the knot group, but of course not vice versa. The base point in the knot group is replaced by the point ∞ in the disc at infinity (but in our case it is now on the knot). At least our 1-cocycles make in this sense some use of the peripheral system. However, we do not know yet if the knot invariants, which are obtained by evaluating the 1-cocycles on canonical loops in M_n, also contain some information from the peripheral system.

The modulus spaces M_n (but without the point ∞) were already introduced in our previous monograph. In this book, we construct new 1-cocycles in M_n. Let us call these 1-cocycles the *combinatorial 1-cocycles*. Hatcher has shown that, e.g., for a long hyperbolic knot K in S^3, its connected component in the moduli space of long knots deformation retracts onto a 2-dimensional torus, generated by two universal loops, known as *Gramain's loop* and the *Fox–Hatcher loop*. Consequently, cocycles of higher dimension would not contain more information in this case and we can restrict our self just to 1-cocycles. There is a third loop because we now consider knots in the solid torus V instead of just long knots. It is defined by pushing in V the string link T (which serves as the closure in V of the n-cable of the long knot) n times through the n-cable. We call this loop $push(T)$. In particular, we can take for T the n-cable of another framed long knot twisted by a cyclic permutation braid. In this way, $push(T)$ induces a *pairing* on the set of all isotopy types of framed knots in 3-space.

Our combinatorial 1-cocycles give invariants for homology classes of oriented loops in M_n and they give knot invariants when we evaluate them on the canonical universally defined loops in the connected components of M_n, namely, Gramain's loop, the Fox–Hatcher loop and the push(T) loop.

Note that there are more general components of M_n, which do not come from the cabling construction, but our combinatorial 1-cocycles are still well defined on them. For example, if we switch just one crossing in the n-cable of K, then the corresponding knot in the solid torus is no longer a closure of an n-cable of a classical knot. The loop $push(T)$ and the Fox–Hatcher loop are no longer defined but Gramain's loop is still well defined, and we can evaluate our combinatorial 1-cocycles on it.

A combinatorial 1-cocycle associates an integer with each Reidemeister III move and the value of the 1-cocycle on a loop is the sum of the integers over all Reidemeister III moves in the loop. The construction of the 1-cocycles is based on singularity theory together with rather complicated combinatorics. The singularity theory is the study of the discriminant Σ of all singular (i.e., non-generic) projections of a (non-singular) smooth oriented knot type in V into the annulus. (This discriminant is very different from Vassiliev's discriminant of singular knots.) This part is completely handled in our previous book and we refer the interested reader to it.

The combinatorics splits into three types of equations to solve, and in the order given in the following, because we consider only regular isotopy (i.e., the projection of the knot into the annulus is always an immersion) and only R III moves contribute to the 1-cocycles:

(a) *the commutation relations*: an R III move with simultaneously another Reidemeister move;
(b) the *positive global tetrahedron equation*: going around a positive quadruple crossing in the moduli space M_n;
(c) the *cube equations*: going around triple crossings where two branches are ordinary tangential in M_n.

The point (b) is by far the hardest part, in analogy to the fact that in the construction of combinatorial 0-cocycles the invariance under Reidemeister III moves is much harder to prove as the invariance under Reidemeister I and II moves. It gives new solutions of a global tetrahedron equation and it could perhaps be useful in other parts of mathematics too (as the theory of integrable systems or string topology).

This book exclusively deals with the construction of the combinatorial 1-cocycles. We have tried to make the definitions as clear as possible so that the reader can use them easily. The proofs of the invariance are rather direct but long and complicated. Therefore, they are separated in a chapter for motivated readers. Calculations of examples by hand are rather tedious and we give only the minimal number of examples but just enough to show that all of our new 1-cocycles are non-trivial. It is a very challenging problem to write computer programs to calculate the resulting knot invariants and to see what can be done with them.

The book is divided into a long introduction, which gives a complete overview of combinatorial 1-cocycles in knot theory and which explains our method and the main ideas, and five other chapters.

Chapter 2 contains the most important results: first lifts of the whole Conway polynomial to four combinatorial 1-cocycles in M_n by using the *distinguished couples of persistent crossings* (a persistent crossing is a crossing which survives if we lift the knot into the n-fold cyclic covering of the solid torus). These 1-cocycles come in couples. *We always need the point at infinity, but the sum of the two 1-cocycles in each couple is a 1-cocycle which does no longer depend on the choice of the point at infinity.*

We give then another lift of the coefficient of z^2 of the Conway polynomial to a combinatorial 1-cocycle in M_n by using this time the *distinguished triples of persistent crossings* and its more sophisticated conjectural generalisation to another lift of the whole Conway polynomial.

Next, we give an essential refinement of all our combinatorial 1-cocycles, including those from our previous monograph.

Finally, we define a refined combinatorial 1-cocycle which uses all crossings at the same time for a cable of a long knot. It depends on three cyclically ordered parameters.

We indicate then as the hypothetical applications of our 1-cocycles:

- *they represent the Conway polynomial as a combinatorial 1-cocycle evaluated on a very simple loop in the moduli space;*
- *they distinguish perhaps orientations of knots;*
- *they detect perhaps all torus knots (including the trivial knot) by invariants which are calculable with polynomial complexity.*

Chapter 3, which is long, contains the proofs.

Chapter 4 contains the examples. We give them in many details to make it easier for the reader to understand the definitions of our 1-cocycles and to make more calculations and applications by his/her own.

For completeness, the last two chapters contain some "forgotten" linear 1-cocycles which should already have been in the previous book, as well as our eclectic 1-cocycle, which was constructed earlier by gluing a 1-cochain from the HOMFLYPT polynomial to a finite type 1-cochain.

I wish to thank Roland van der Veen for interesting discussions and for programming some of the 1-cocycles, which turned out to be very useful in order to show that they are non-trivial. Finally, let me mention that without Séverine, who has encouraged me over all these years and who has in particular created all the figures, this book wouldn't exist.

About the Author

Thomas Fiedler is Professor at the Institut de Mathématiques de Toulouse, Université Paul Sabatier in France since 1992. He has studied Mathematics at the Leningrad State University and graduated under the direction of V.A. Rokhlin in 1977. After many years of work about the topology of real algebraic projective planar curves, he has become interested in knot theory, and he is now the author of three monographs in this field. Since 2005, he has been working on a new approach in knot theory, namely, the construction of combinatorial 1-cocycles in spaces of knot diagrams. They are based on solutions of a global tetrahedron equation, which enables us to study classical knots by using combinatorial analysis instead of representation theory. Since 2020, there has been a fruitful collaboration with Roland van der Veen from the University of Groningen, in order to create a computer program for calculating the new 1-cocycles on canonical loops in spaces of diagrams.

Contents

List of Figures

Chapter 1

Introduction

The topology of moduli spaces of classical knots and of long knots was much studied in Refs. [5–8,25]. In particular, there is a complete description of the homotopy type and of the homology groups (even with an additional structure of certain Gerstenhaber–Poisson algebras) for the space of long knots in $\mathbb{R}^3$ (the same space, which was studied by Vassiliev using singularity theory) in the subsequent Refs. [6–8,25]. Vassiliev [46] had constructed certain universally defined 0-dimensional cohomology classes. They give the finite type invariants of long knots (or equivalently of compact knots in the 3-sphere) when they are evaluated on the 0-dimensional homology classes of the disconnected space. It is known that all finite type knot invariants, also called Vassiliev–Goussarov invariants, have diagrammatic formulas, which allow the computation of the invariant from an arbitrary generic knot diagram [24]. Our goal in Ref. [16] was the construction of diagrammatic formulas for 1-dimensional cohomology classes of the space of knots in the solid torus, which give knot invariants when they are applied to universally defined loops, represented by generic 1-parameter families of diagrams for knots in the solid torus and in particular for cables of long knots.

The present book is the sequel of the monograph. We study the global tetrahedron equation in a more systematic way and as a consequence we obtain new much more elaborated combinatorial 1-cocycles. In particular, give four different lifts of the Conway polynomial to combinatorial 1-cocycles in the moduli space of diagrams M_n.

We fix a linear projection $pr : \mathbb{C} \times \mathbb{R} \to \mathbb{C}$ of the 3-space into the plane, and knots in 3-space can be given now by knot diagrams, i.e., a smoothly embedded oriented circle in 3-space together with its projection into the plane. It is well known that oriented knot types are in 1–1 correspondence with knot types of oriented long knots. We close now the long knot with the 1-braid to a knot in the standard embedded solid torus $V^3 \subset \mathbb{R}^3$. Moreover, a (black board) framed long knot K' can be replaced by its parallel n-cable, called nK' (with the same orientation of all strands), which we close to a knot in V^3 by a cyclic permutation braid or more generally by any fixed n-component string link T but which induces a cyclic permutation of its end points. We denote the resulting knot in V^3 by $K = T \cup nK'$. The projection into the plane becomes now a projection into the annulus $pr : \mathbb{C}^* \times \mathbb{R} \to \mathbb{C}^*$. We chose as generator of $H_1(V^3)$ the class which is represented by the closure of the oriented 1-braid. Hence, the knot in V^3, which is obtained from the n-cable, represents the homology class $n \in H_1(V^3)$. We consider the infinite dimensional space M_n of all diagrams of knots K in V^3, which represent the homology class n such that the projection $pr : K \to \mathbb{C}^*$ is an immersion and $pr(K)$ does not contain negative loops. Moreover, we fix a point in a canonical way, namely, on the lowest strand on the intersection of K with the disc at infinity (compare the figures), called ∞, and we add a new condition, called the *separation condition*: a crossing cannot move over the point ∞ at the same moment as a Reidemeister III move happens somewhere in the diagram (i.e., Reidemeister III moves and sliding crossings over the point ∞ are separated, in particular, they are not allowed to commute). M_n is called the *moduli space of long regular knot diagrams without negative loops in the solid torus*. The knot types up to regular isotopy in 3-space correspond to the connected components of M_1. Given a generic knot diagram $K \subset V^3$, we consider the oriented curve $pr(K)$ in the annulus. A *loop in $pr(K)$* is a piecewise smoothly oriented immersed circle in $pr(K)$ which respects the orientation of $pr(K)$. In other words, we go along $pr(K)$ following its orientation, and at a double point, we are allowed to switch perhaps to the other branch but still following the orientation of $pr(K)$. Naturally, a loop in $pr(K)$ is called *negative* (respectively, *positive*) if it represents a negative (respectively, positive) homology class in $H_1(V^3)$. One easily sees that knots which arise as cables of long knots by the above construction never contain negative loops. This will turn out to be very important.

Indeed, if all loops are positive, then the knot is a closed braid and "quantum 1-cocycles" can be defined directly (see Ref. [16]).

If all loops are non-negative, then "quantum 1-cocycles" can be defined by lifting to 1-cocycles the finite type invariants which can be extracted from quantum polynomials (this is started in this monograph).

If there are negative loops as well, then our method breaks down, at least for the moment.

Victor Vassiliev has started the combinatorial study of the space of long knots in *Combinatorial formulas of cohomology of knot spaces* [47] by studying the simplicial resolution of the discriminant of long singular knots in $\mathbb{R}^3$. Our approach is very different because it is based on the study of another discriminant, namely, the discriminant of non-generic diagrams of knots in the solid torus [16]. There is also a big difference for H_1 of the moduli spaces. Let $K_1 \# K_2$ be the long knot which corresponds to the connected sum of two different knots. In the moduli space of long knots, we have a loop which consists of pushing the knot K_1 from the left to the right through the knot K_2 and then pushing K_2 through K_1 from the left to the right too [23]. In M_n, we can just push a knot $T \cup nK_1$ through nK_2 from the left to the right and then slide without any Reidemeister moves $T \cup nK_1$ again in its initial position along the remaining part of the solid torus. This is already a loop. We show that our 1-cocycles define in this way non-trivial *pairings* on $H_0(M_n) \times H_0(M_1)$ for $n > 1$.

Why do we need to construct 1-cocycles? Classical knots can be transformed into knots in the solid torus. The moduli space of knots in the solid torus is essentially the only moduli space of knots in 3-manifolds (and which are not contained in a 3-ball) with infinite H_1, compare Ref. [16] (the moduli space of a long hyperbolic knot deformation retracts onto a 2-dimensional torus [25] and hence the study of H_1 will be sufficient). We have to use this because so far the traditional study of H_0 (which has given plenty of invariants) is not enough to distinguish all classical knots.

Indeed, 30 years after their discovery, there are still not known any finite type invariants which could distinguish the orientations of classical knots. An observation of Greg Kuperberg [33] says that if finite type invariants cannot distinguish the orientations of knots, then they fail even to distinguish non-oriented knots as well.

But why do we need to construct 1-cocycles in a combinatorial way? Of course, it would be of great importance for better understanding (and perhaps in order to make connections with string theory) to construct differential 1-forms on the moduli space M_n, which depend on $a \in H_1(V^3; \mathbb{R})$ as parameter and which represent the same cohomology classes as our combinatorial 1-cocycles if the parameter is in $H_1(V^3; \mathbb{Z})$. But they do not exist yet and we have to put up with the difficult combinatorial approach.

Combinatorial integer 0-cocycles are usually called *Gauss diagram formulas*, compare Refs. [15,40]. They correspond to finite type invariants and are solutions of the 4T- and 1T-relations. Dror Bar-Natan has shown in *On the Vassiliev knot invariants* [2] that such solutions can be constructed systematically by using the representation theory of Lie algebras. Arnaud Mortier has constructed for 1-cocycles of finite type for long knots the analogue of the Kontsevich integral in *A Kontsevich integral of order 1* [36]. The 4T-relations are now replaced by three 16T- and three 28T-relations and 4×4T-relations! There is actually no representation theory which could help to construct such solutions. The reason for this is simple. The well-known representation theory related to the tetrahedron equation, see, e.g., Ref. [30], is as usual of a local nature. But H_1 of the moduli space of (closed) non-satellite knots in $\mathbb{R}^3$ is only torsion (in contrast to H_0), see Refs. [7,8,25], and hence, all integer 1-cocycles from local solutions of the tetrahedron equation are trivial (in contrast to the local solutions from the Yang–Baxter equation, see, e.g., Ref. [31])! We construct therefore in a combinatorial way solutions of the *global tetrahedron equation*, i.e., the contribution of an R III move depends on the whole knot in the solid torus and not only on the local picture of the move. This is an equation which is much more complicated as the well-known Yang–Baxter equation.

It seems that before [16] only the moduli space of long knots was studied. The Teiblum–Turchin 1-cocycle v_3^1 was the only known 1-cocycle for long knots which represents a non-trivial cohomology class and which has (probably) an explicit diagrammatic formula for its computation. The Teiblum–Turchin 1-cocycle is an integer-valued 1-cocycle of degree 3 in the sense of Vassiliev's theory [46]. Its reduction mod 2 had the first diagrammatic description of a 1-cocycle, see

Refs. [45,47]. Sakai has defined an $\mathbb{R}$ valued version of the Teiblum–Turchin 1-cocycle via configuration space integrals [42]. We have found a very complicated formula for an integer extension of v_3^1 mod 2 in Ref. [18] by combining the HOMFLYPT polynomial with a finite-type 1-cocycle. For completeness, we repeat its construction in Chapter 6. The most beautiful diagrammatic formula for an integer-valued 1-cocycle for long knots which extends v_3^1 mod 2 and which probably coincides with v_3^1 was found by Mortier [34,35]. Budney has defined another integer-valued 1-cocycle on the space of long knots in Ref. [7], Propositions 6.1 and 6.3, using Gramain's loop. The value of his 1-cocycle on Gramain's loop (see the following for a definition) for a knot K is n if K is the connected sum of exactly n non-trivial knots. But it is not clear how to represent it by a diagrammatic 1-cocycle.

It is well known that if a knot K in the 3-sphere is not a satellite, then its topological moduli space is a $K(\pi, 1)$ with π a finite group (in fact $\pi = Aut(\pi_1(S^3 \setminus K), \partial)$, where ∂ is the peripheral system of the knot K, see Refs. [26,27,48]). Consequently, in this case, there can't exist any non-trivial 1-cocycles with values in a torsion free module. However, it is also well known that the components of the moduli space of knots in the 3-sphere are in a natural 1–1 correspondence with the components of the moduli space of long knots in 3-space. Hatcher [25] has proven that for long knots in 3-space, the situation is much better. There are always two canonical non-trivial loops in the component of the topological moduli space of a long knot K if it is not the trivial knot: Gramain's loop, denoted by $rot(K)$, and the Fox–Hatcher loop, denoted by $fh(K)$. *Gramain's loop* is induced by the rotation of the 3-space around the long axis of the long knot [23]. *Fox–Hatcher's loop* is defined as follows: one puts a pearl (i.e., a small 3-ball B) on the closure of the long framed knot K in the 3-sphere. The part of K in $S^3 \setminus B$ is a long knot. Pushing B once along the knot with respecting the framing induces the Fox–Hatcher loop, see Refs. [20,25]. The homology class of $rot(K)$ does not depend on the framing of K, and changing the framing of K adds multiples of $rot(K)$ to $fh(K)$. Note that the Fox–Hatcher loop has a canonical orientation induced by the orientation of the long knot. The same loops are still well defined and non-trivial for those n-string links which are n-cables of a framed non-trivial long knot. It follows from the results of Hatcher [25] and Budney [8] that these two loops are

linearly dependent in the rational homology if and only if the knot
is a torus knot. Moreover, Hatcher has shown that the topological
moduli space of a long hyperbolic knot deformation retracts onto a
2-dimensional torus. Hence, it follows from Künneth's formula that
it is sufficient to construct just 1-cocycles in this case.

Let us try to make it completely clear to the reader: the homology
of the space of long knots and hence of M_n depends very strongly on
the component of the space. But luckily, each component contains
two canonical 1-homology classes, namely, the class represented by
Gramain's loop and the class represented by the Fox–Hatcher loop
(which are dependent or even trivial only in very particular cases).
We can represent these loops and their natural generalisations in M_n
by 1-parameter families of diagrams. Moreover, we have in M_n in
addition the loop $push(T)$, as already explained. We are now interested in those 1-cocycles in M_n which are universally defined (i.e.,
independent of the component of the space) and which can be calculated from the 1-parameter family of diagrams alone.

We will explain now our method in some detail. We assume
that all our finite dimensional spaces and manifolds are smooth and
oriented.

Definition 1.1 *We fix a orthogonal projection* $pr : \mathbb{C} \times \mathbb{R} \to \mathbb{C}$
together with standard coordinates (x, y, z) *of* $\mathbb{C} \times \mathbb{R}$. *A long knot* K'
is an oriented smoothly embedded copy of $\mathbb{R}$ *in* $\mathbb{C} \times \mathbb{R}$ *which coincides with the real x-axis in* $\mathbb{C} \times 0$ *outside a compact set. A parallel
n-cable of a framed long knot* nK' *is a n-component link with fixed
end points where each component is parallel to the framed long knot
with respect to the blackboard framing given by pr (the z-coordinate)
and with the same orientation on each component. An n-string link
T is an n-component link with fixed end points where each component
is parallel to a long knot in* $\mathbb{C} \times 0$ *outside some compact set.*

We put the point ∞ *on the lowest component of* nK' *or* T *with
respect to the y-coordinate at infinity.*

We cut now the string link nK' with a very big 3-ball B^3. The
end points of the string link are, respectively, in two big discs (called
the disc at infinity) which we glue together in order to obtain a solid
torus V. We chose a string link T which we glue to nK' near to the
disc at infinity at $x = -\infty$ and such that $T \cup nK'$ is an oriented knot
$K \subset V$.

It turns out that M_n is the right space in order to study classical knots. Evidently, $K = T \cup nK'$ does not contain any negative loops. Moreover, if a framed long knot K'' is (framed) isotopic to K', then the corresponding knots $K_1 = T \cup nK'$ and $K_2 = T \cup nK''$ are in the same component of M_n (compare, e.g., Ref. [15]). Consequently, each 1-cocycle of M_n evaluated on a universally defined loop in M_n (i.e., the definition of the loop does not depend on the component of M_n) gives an invariant of classical knots. It is often convenient for calculations to represent a long knot K' as a closed braid with just one strand opened to go to infinity.

But our 1-cocycles are not only useful in order to define knot invariants. A well-known question in 3-dimensional knot theory is as follows: Given a complicated knot diagram, is this the unknot, or more generally, do two given knot diagrams represent the same knot type?

An analogue of this question on the next level is as follows: Given two loops in the moduli space M_n by complicated movies of diagrams, are these loops homotopic? (There isn't known any algorithm in order to simplify the loops.) However, our 1-cocycles allow one sometimes to give a negative answer to this question, even without recognising the knot type.

For the convenience of the reader, we remind now the basic technical result and our strategy developed in Ref. [16].

It follows from Thom–Mather singularity theory that each component of the infinite dimensional space M_n has a natural *stratification with respect to pr*:

$$M_n = \Sigma^{(0)} \cup \Sigma^{(1)} \cup \Sigma^{(2)} \cup \Sigma^{(3)} \cup \Sigma^{(4)} \cdots$$

Here, $\Sigma^{(i)}$ denotes the union of all strata of codimension i.

The strata of codimension 0 correspond to the usual generic *diagrams of knots*, i.e., all singularities in the projection are ordinary double points. So, our *discriminant* is the complement of $\Sigma^{(0)}$ in M_n. Note that this discriminant of non-generic diagrams is very different from Vassiliev's discriminant of singular knots [46].

The three types of strata of codimension 1 correspond to the *Reidemeister moves*, i.e., non-generic diagrams which have exactly one ordinary triple point, denoted by $\Sigma^{(1)}_{\text{tri}}$, or one ordinary self-tangency, denoted by $\Sigma^{(1)}_{\text{tan}}$, or one ordinary cusp, denoted by $\Sigma^{(1)}_{\text{cusp}}$, in the projection pr. We call the triple point together with the under–over information (i.e., its embedded resolution) a *triple crossing*. We distinguish self-tangencies for which the orientation of the two tangents coincide, called $\Sigma_{\text{tan}+}^{(1)}$, from those for which the orientations of the tangents are opposite, called $\Sigma_{\text{tan}-}^{(1)}$.

Proposition 1.1 (see Ref. [16]) *There are exactly six types of strata of codimension* 2. *They correspond to non-generic diagrams which have exactly either*

(1) *one ordinary quadruple point, denoted by* $\Sigma_{\text{quad}}^{(2)}$;

(2) *one ordinary self-tangency with a transverse branch passing through the tangent point, denoted by* $\Sigma_{\text{trans}-\text{self}}^{(2)}$;

(3) *one ordinary self-tangency in an ordinary flex* $(x = y^3)$, *denoted by* $\Sigma_{\text{self}-\text{flex}}^{(2)}$;

(4) *two singularities of codimension* 1 *in disjoint small discs* (*this corresponds to the transverse intersection of two strata from* $\Sigma^{(1)}$, *i.e., two simultaneous Reidemeister moves at different places of the diagram*);

(5) *one ordinary cusp* $(x^2 = y^3)$ *with a transverse branch passing through the cusp, denoted by* $\Sigma_{\text{trans}-\text{cusp}}^{(2)}$;

(6) *one degenerate cusp, locally given by* $x^2 = y^5$, *denoted by* $\Sigma_{\text{cusp}-\text{deg}}^{(2)}$;

We show these strata in Fig. 1.1.

Our strategy is the following: for an oriented generic loop in M_n, we associate an integer to the intersection with each stratum in $\Sigma_{\text{tri}}^{(1)}$, i.e., to each Reidemeister III move, and we sum up over all moves in the loop.

In order to show that our 1-cochains are 1-cocycles, we have to prove that the sum is 0 for each meridian of strata in $\Sigma^{(2)}$. This is very complex, but we use strata from $\Sigma^{(3)}$ in order to reduce the proof to a few strata in $\Sigma^{(2)}$. It follows that our sum is invariant under generic homotopies of loops in M_n. But it takes its values in an abelian ring and hence it is a 1-cocycle. Showing that the 1-cocycle is 0 on the meridians of $\Sigma_{\text{quad}}^{(2)}$ is by far the hardest part. This corresponds to finding a new solution of the *tetrahedron equation*. Consider four oriented straight lines which form a braid such that the intersection

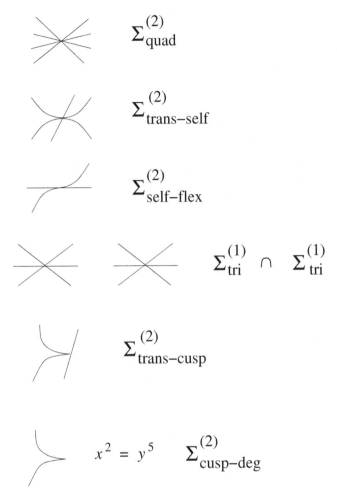

Figure 1.1: The strata of codimension 2 of the discriminant of non-generic projections.

of their projection into $\mathbb{C}$ consists of a single point. We call this an *ordinary quadruple crossing*. After a generic perturbation of the four lines, we will see now exactly six ordinary crossings. We assume that all six crossings are positive, and we call the corresponding quadruple crossing a *positive quadruple crossing*. Quadruple crossings form smooth strata of codimension 2 in the topological moduli space of lines in 3-space which is equipped with a fixed projection pr. Each generic point in such a stratum is adjacent to exactly eight smooth strata of codimension 1. Each of them corresponds to configurations of lines which have exactly one ordinary triple crossing besides the remaining ordinary crossings. We number the lines from 1 to 4 from the lowest to the highest (with respect to the projection pr). The eight strata of triple crossings glue pairwise together to form four smooth strata which intersect pairwise transversely in the stratum of the quadruple crossing, see, e.g., Ref. [19]. The strata of triple crossings are determined by the names of the three lines which give the triple crossing. For shorter writing, we give them names from P_1 to P_4 and $\bar{P}_1$ to $\bar{P}_4$ for the corresponding stratum on the other side of the quadruple crossing. We show the intersection of a normal 2-disc of the stratum of codimension 2 of a positive quadruple crossing with the strata of codimension 1 in Fig. 1.2. The strata of codimension 1 have a natural coorientation, compare the next section.

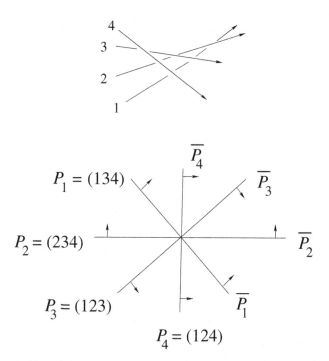

Figure 1.2:   The intersection of a normal 2-disc of a positive quadruple crossing with the strata of triple crossings.

We could interpret the six ordinary crossings as the edges of a tetrahedron and the four triple crossings likewise as the vertices or the 2-faces of the tetrahedron.

For the classical tetrahedron equation, one associates to each stratum P_i, i.e., to each vertex or equivalently to each 2-face of the tetrahedron, some operator (or some R-matrix) which depends only on the names of the three lines and to each stratum $\bar{P}_i$ the inverse operator. The tetrahedron equation says now that if we go along the meridian, then the composition of these operators is equal to the identity. Note that in the literature, see, e.g., Ref. [30], one considers planar configurations of lines. But this is of course equivalent to our situation because all crossings are positive and hence the lift of the lines into 3-space is determined by the planar picture. Moreover, each move of the lines in the plane which preserves the transversality lifts to an isotopy of the lines in 3-space. The tetrahedron equation has many solutions; the first one was found by Zamolodchikov, see, e.g., Ref. [30].

However, the solutions of the classical tetrahedron equation are not well adapted in order to construct 1-cocycles for moduli spaces of knots. A local solution of the tetrahedron equation is of no use for us because as already pointed out there are no integer-valued 1-cocycles for knots in the 3-sphere. We have to replace them by long knots or more generally by points in M_n. For knots in the solid torus V, we can now associate with each crossing in the diagram a winding number (i.e., a non-negative homology class in $H_1(V)$) in a canonical way. Therefore, we have to consider six different positive tetrahedron equations, corresponding to the six different abstract closures of the four lines to a circle, and in each of the six cases, we have to consider all possible winding numbers of the six crossings. Moreover, in each case, we have to consider the four possibilities for the point ∞. We call this the *positive global tetrahedron equations*. One easily sees that there are exactly *forty eight* local types of quadruple crossings (analogue to the eight local types of triple crossings). We study the relations of the local types in what we call the *cube equations* (compare Ref. [16]).

The construction of the 1-cocycles is complex but the outcomes are rather beautiful formulas. Surprisingly, it becomes now really essential that the projection of $T \cup nK'$ into the annulus contains no loops which represent a negative homology class that there is a fixed point at infinity and that the long knots are framed. (Usually, the invariance of quantum knot invariants under the R I moves is just a matter of normalisation. This is no longer the case for 1-cocycles on 1-parameter families of diagrams: the place of the R I move in the diagram and the moment in the 1-parameter family now become very important.)

It follows from Proposition 1.1 that in order to obtain combinatorial 1-cocycles, there are only three types of equations to solve, and in the order given in the following, because we consider only regular isotopy and only R III moves contribute to the 1-cocycle (compare Ref. [16]):

(a) (corresponds to (4) in Proposition 1.1) *the commutation relations*: an R III move with simultaneously another Reidemeister move;

(b) (corresponds to (1)) *the positive global tetrahedron equation*: going around a positive quadruple crossing in the moduli space;

(c) (corresponds to (2)) *the cube equations*: going around triple crossings where two branches are ordinary tangential in the moduli space.

To make it easier for the reader to understand our method, we restrict ourself here on a particular case.

So far, all our 1-cocycles constructed in Ref. [16] have used *linear weights* for the contributions of R III moves to the 1-cocycles, i.e., besides the triangle (which corresponds to the move in the Gauss diagram) we consider just the position of individual arrows (which correspond to the crossings) with respect to the triangle. More generally, the construction of combinatorial 1-cocycles needs as an input *Gauss diagram formulas for finite type invariants of long knots*, see, e.g., Refs. [11–13,15,24,40]. In Ref. [17], we have made use of the Polyak–Viro formula for $v_2(K)$ of long knots shown on the left in Fig. 1.3.

We want to construct for each natural number k *weights* $W_{2k}(p)$ of order $2k$ for each R III move p which is defined by using $2k$-tuples of crossings (i.e., arrows in the Gauss diagram which always go from the undercross to the overcross, compare Ref. [16]) outside the R III move p (i.e., arrows not of the triangle in the Gauss diagram which corresponds to the R III move). But these $2k$-tuples have to be

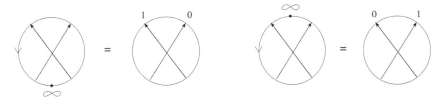

Figure 1.3: Polyak–Viro formulas for $v_2(K)$ of long knots.

related to the R III move p. The commutation relations (a) force the weights $W_{2k}(p)$ to be knot invariants if we consider all 2k-tuples and not only those which are related to the move p. R I moves force the weights not to contain isolated arrows with homological markings 0 or n (see the following). R II moves force that each crossing of the weights contributes with its sign (writhe). Consequently, it is then sufficient to show that the weight $W_{2k}(p)$ are invariant under a simultaneous R III move with the move p.

An invariant of long knots which is given by a Gauss diagram formula, which uses only 2k-tuples of arrows, is an invariant of order $2k$. For example, for $k = 1$, it can be defined by the beautiful formulas of Michael Polyak and Oleg Viro for $v_2(K)$ of long knots [40], see Fig. 1.3. The point on the circle corresponds to the point at infinity on the knot.

Definition 1.2 *For a crossing q, we call D_q^+ the knot which is obtained by smoothing the crossing q from the undercross to the overcross (and the remaining knot is called D_q^-), see Fig. 1.4. The homological marking of q is the homology class in $\mathbb{Z} \cong H_1(V^3)$ represented by D_q^+ (compare Ref. [16]), i.e., by the knot which corresponds to the arc of the circle from the overcross to the undercross.*

Consequently, the point at infinity is in D_q^+ for long knots if and only if the homological marking $[q] = 1$.

The sign (or writhe) of a crossing q is denoted as usual by $w(q)$. An ordinary crossing q is called an n-crossing, respectively, 0-crossing, if its homological marking $[q] = n$, respectively, $[q] = 0$.

Definition 1.3 *Let $T \cup nK'$ be a knot in M_n. We consider the cyclic n-fold covering of V^3 and we lift $T \cup nK'$ to a knot $\tilde{K}$ in the covering. $\tilde{K}$ can be naturally identified with a long knot (∞ is on $\tilde{K}$) and we call it the underlying long knot for $T \cup nK'$.*

Figure 1.4:   The two ordered knot diagrams associated with a crossing q.

The crossings of $\tilde{K}$ are called the *persistent crossings*. These are exactly the crossings of $T \cup nK'$ with the markings 0 and n.

Proposition 1.2 *Let G be a Gauss diagram formula for long knots which defines a knot invariant. Let $K \subset V^3$ be a knot which belongs to M_n (i.e., no negative loops). We keep the markings 0 in G but we replace each marking 1 by the marking n. Then, the resulting Gauss diagram formula defines a knot invariant for K in M_n.*

Proof.   We consider the cyclic n-fold covering V^3 over V^3, and we lift K to the underlying long knot $\tilde{K}$ in it. Crossings with marking 0 stay crossings with marking 0, crossings with marking n become crossings with marking 1 and all other crossings disappear. The fact that G is an invariant for long knots implies the result.   □

Note that the result is no longer true if negative loops are allowed as well, already for the invariant of order two (because more markings appear for $\tilde{K}$), see Fig. 1.5.

Let's come back to the formula for v_2, which is important in this book.

We will use for $W_2(p)$ the following configuration, which corresponds to the Polyak–Viro formula on the left in Fig. 1.3 (where as usual we take the product of the signs of the two crossings) and which we denote shortly by $(n, 0)$, see Fig. 1.6.

To each Reidemeister move of type III corresponds a diagram with a *triple crossing* p: three branches of the knot (the highest, middle and lowest with respect to the projection $pr : \mathbb{C}^* \times \mathbb{R} \to \mathbb{C}^*$) have a common point in the projection into the plane. A small perturbation of the triple crossing leads to an ordinary diagram with three crossings near $pr(p)$.

Definition 1.4 *We call the crossing between the highest and the lowest branch of the triple crossing p the* distinguished crossing *of p and we denote it by d ("d" stands for distinguished). The crossing*

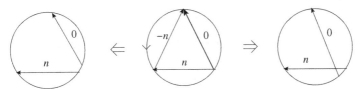

Figure 1.5: The Polyak–Viro formula for $v_2(K)$ is not true if there are negative loops.

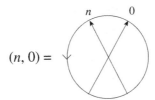

Figure 1.6: Configuration (or arrow diagram) $(n, 0)$.

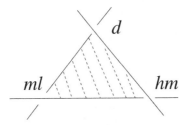

Figure 1.7: The names of the crossings in an R III move.

between the highest branch and the middle branch is denoted by hm and that of the middle branch and the lowest branch is denoted by ml, compare Fig. 1.7. For better visualisation, we draw the crossing d always with a thicker arrow. Moreover, we give the same name to the adjacent arc of the crossing in the circle, compare Fig. 1.8.

Let p be a triple crossing. The triangle in the Gauss diagram cuts the circle into three open arcs, which we denote by the name of the adjacent crossing in the triangle. Moreover, we indicate on which side of the crossing is the arc: in Fig. 1.7, the arcs correspond to d^+, hm^- and ml^- (compare Fig. 1.4).

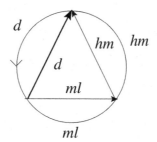

Figure 1.8:   The names of the arcs in an R III move.

Definition 1.5 *Let p be a triple crossing (which depends on an integer parameter $0 < a < n$) such that hm is a persistent crossing (i.e., it has the marking 0 or n). An ordinary crossing q of the diagram is called an f-crossing for p if $[q] = n$ and the foot of q is in the open arc hm^+.*

An ordinary crossing q of the diagram is called an r-crossing for p if $[q] = 0$ and the foot of q is in the open arc hm^+.

Let us consider, for example, the case $k = 1$. We make a connection of the weight with the triple crossing p by defining the *weight of p of order 2* as $W_2(p) = \sum(n, 0)$, where the sum is taken only over all f-crossings for p (and no restrictions on the foots of the 0-crossings here).

This is what we call a *distinguished couple of persistent crossings*: the couple of persistent crossings are the crossing hm and the f-crossing.

It turns out that the weight $W_2(p)$ solves the commutation relations (a), but it does not solve the positive global tetrahedron equation (b). The eight strata of R III moves in the meridian of a positive quadruple crossing in the moduli space come in pairs with different signs: P_i and $\bar{P}_i$ (compare Ref. [16]). They can differ by n-crossings for which the foot has slide over the head or the foot of a crossing in the triangle of P_i or $\bar{P}_i$, and hence, sometimes these crossings do no longer contribute to $W_2(p)$. The key point is that we consider only R III moves of a very particular global type (see Section 2.2 in Chapter 2). It turns out then that an n-crossing which changes from P_i to $\bar{P}_i$ the arc of its foot is always a crossing hm for another R III move P_j and $\bar{P}_j$ of the same particular global type in the meridian of the quadruple crossing. We use this to define a weight of order 1 for P_j and $\bar{P}_j$, which we multiply by a certain refined linking number (which depends on the integer parameter a) and which will contribute to the 1-cocycle too. The weight of order 1 is essentially the same for P_j and $\bar{P}_j$ but the refined linking numbers are different. This is really complex but leads finally to a solution of the positive global tetrahedron equation (b). The solution of the cube equations (c) is then relatively easy and leads just to some simple linear correction terms in the 1-cocycle.

Let us mention that the construction of the 1-cocycle completely breaks down for the parameters $a = 0$ and $a = n$ (and hence we cannot apply it for $n = 1$, and in particular it becomes essential again that knots have to be framed because the isotopy class of cables depends on the framing). We will give the precise result in the next chapter.

Note that we have made several choices. The two Polyak–Viro formulas for v_2 lead to four different "dual" 1-cocycles: head instead of foot for n-crossings, foot or head in ml^+ instead of hm^+. If we replace a Gauss diagram formula by its mirror image, then in Definition 1.5 foots have to be replaced by heads. Also the group $\mathbb{Z}/2\mathbb{Z} \times \mathbb{Z}/2\mathbb{Z}$ acts naturally on the moduli spaces, generated by orientation reversing of K together with the hyper elliptic involution of the solid torus and by taking the mirror image. Some of the corresponding 1-cocycles correspond simply to the different choices in the construction.

In this book, we will make use of the generalisation of the Polyak–Viro formula by Chmutov–Khoury–Rossi [13] and Brandenbursky [4] to Gauss diagram formulas for all coefficients of the Conway polynomial.

For the convenience of the reader, we repeat these formulas here, compare Ref. [13].

Let $k \in \mathbb{N}$ be fixed and let A_{2k} be an arrow diagram (i.e., an abstract Gauss diagram without signs on the arrows, we call it often a configuration) with one (oriented) circle, $2k$ arrows and a base point. Following Chmutov–Khoury–Rossi, A_{2k} is called *ascending one-component* if by going along the oriented circle starting from the base point and each time jumping along the arrow we meet each arrow first at its foot and the traveling meets the whole circle.

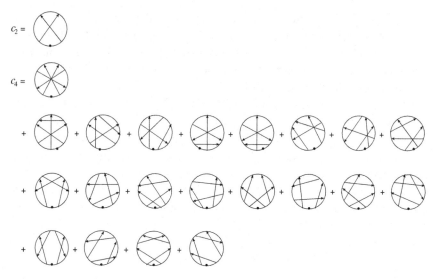

Figure 1.9:   The Chmutov–Khoury–Rossi formulas for $k = 1$ and $k = 2$.

The Gauss diagram formula C_{2k} is simply the sum of *all* ascending one-component arrow diagrams A_{2k}. Each arrow diagram enters of course with the sign which is the product of all signs of crossings in it. We show as examples C_2 and C_4 in Fig. 1.9. By definition, $C_0 = 1$.

Chmutov, Khoury and Rossi have proven by using skein relations that C_{2k} evaluated on a long knot diagram K is the coefficient of z^{2k} in the Conway polynomial of K, i.e., $\nabla_K(z) = \sum_{k=0} C_{2k} z^{2k}$. As is well known, this sum is finite and the highest degree of z can be bounded by using the genus of the knot, see, e.g., Ref. [9].

Brandenbursky [4] has interpreted C_{2k} evaluated on a long knot diagram K as the algebraic number of certain surfaces with one boundary component related to K and of Euler characteristics $1 - 2k$ (in the spirit of Gromov–Witten invariants), and he has given a direct proof that it is invariant under the Reidemeister moves (and in particular, there is a natural $1 - 1$ correspondence of contributing configurations before and after the R III move). Hence, the formula C_{2k} even gives an invariant for virtual long knots (this is very important for us in the proof, even if our 1-cocycles are only well defined for real knots, because each loop for virtual or welded knots would contain forbidden moves, see Ref. [32] for the definition of forbidden moves).

The Chmutov–Khoury–Rossi–Brandenbursky formulas have two properties, which are so important that we give names to them as follows:

First–last foot property: If we travel on the oriented circle starting from the point at infinity, then we meet in each configuration A_{2k} always first the foot of an n-crossing and we meet last the foot of a 0-crossing (this is an immediate consequence of their definition). We call this correspondingly the *first n-crossing* and the *last 0-crossing* in each configuration.

Preserving intersection property: There is a 1–1 correspondence of contributing configurations A_{2k} before and after an R III move. If all crossings in the R III move are positive and a couple of intersecting crossings in a configuration (i.e., the arrows intersect in the Gauss diagram) contribute before the move (no matter the direction of the move), then it is replaced by exactly one different couple but also with intersecting crossings in a configuration after the move (this is a consequence of Brandenbursky's proof [4]). Consequently, the corresponding analogue is true as well for couples of non-intersecting crossings. Moreover, if all three crossings of the triple crossing contribute together before the R III move, then they contribute together also after the R III move.

But note that it can happen that a contributing couple of two 0-crossings in A_{2k} is replaced by a contributing couple of a 0-crossing with an n-crossing in another A_{2k} when we perform an R III move. We show an example in Fig. 1.10 (the two configurations correspond to the second and the fourth configurations in the third line of Fig. 1.9).

We use the first–last foot property to construct our 1-cochains, and we use the preserving intersection property in order to show that they are in fact 1-cocycles.

Remark 1.1 *Note also that we haven't used the point ∞ in the construction of the 1-cocycle from the Polyak–Viro formula. This comes from the fact that in this case the homological markings determine the arc which contains ∞. This already is no longer true for the invariant of order 4 (i.e., for $k = 2$): the point ∞ is not determined by the homological markings exactly for the first diagram in the last line of Fig. 1.9. This comes from the fact that the diagram is not*

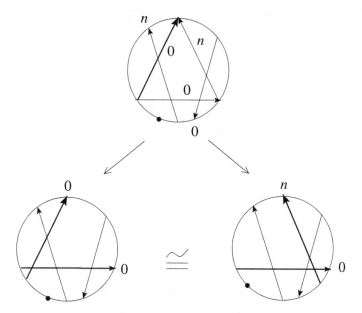

Figure 1.10:   A couple $(0,0)$ is replaced by a couple $(n,0)$.

connected (if we forget about the circle) and that it has a symmetry. In this case, the point ∞ contains additional information.

We lift now each coefficient of the Conway polynomial C_{2k} to a 1-cocycle by using the distinguished couples of persistent crossings and certain types of triple crossings p.

Definition 1.6 *The weight $W_{2k}(p)$ is defined as the sum over all those arrow diagrams A_{2k} for which the first foot (of necessarily an n-crossing) is in the open arc hm^+, in other words it is an f-crossing.*

Next, we lift each coefficient of the Conway polynomial C_{2k} to a 1-cochain by using the distinguished triples of persistent crossings and more types of triple crossings p.

Definition 1.7 *The weight $\bar{W}_{2k}(p)$ is defined as the sum over all those arrow diagrams A_{2k} for which the first foot (of necessarily an n-crossing) is in the closure of hm^+ and the last foot (of necessarily a 0-crossing) is in the closure of hm^+ too.*

We show that it will give rise to a 1-cocycle at least for $k = 1$.

The *distinguished triple of persistent crossings* are the first n-crossing and the last 0-crossing in A_{2k} and the n-crossing hm. Note that the n-crossing hm itself is not necessarily a crossing from A_{2k}, but the foots of the first n-crossing and of the last 0-crossing in A_{2k} have to be in the closure of hm^+. Hence, we have mutual relations between all three persistent crossings, and there is a relation with the two a-crossings in the R III move too.

It is amazing that these two sorts of 1-cocycles (from distinguished couples of persistent crossings and from distinguished triples of persistent crossings) come from completely different sorts of solutions of the global tetrahedron equation.

In order to explain this, we have to consider the global types of R III moves and of the tetrahedron more closely.

A Reidemeister III move p corresponds to a triangle in the Gauss diagram. The *global type of a Reidemeister III move* is now shown in Fig. 1.11, where $m + h$, $m + h - [K]$, m and h are the homological markings of the corresponding arrows. Here, $[K] = n$ is the homology class represented by the knot K. Moreover, we indicate whether the arrow ml in the triangle goes to the left, denoted by l, or it goes to the right, denoted by r. Note that two of the markings determine always the third marking.

It is convenient to encode the global type of an R III move in the following way: $r([d], [hm], [ml])$ and, respectively, $l([d], [hm], [ml])$. (Remember that $[.]$ denotes the homological marking of the crossing.)

Note that the persistent crossings of nK determine the long knot K. We use the homological markings of the remaining crossings as parameters.

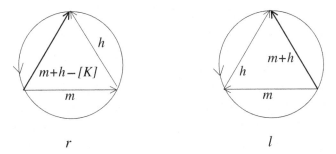

$$r \qquad\qquad l$$

Figure 1.11: The global types of R III moves for knots in the solid torus.

We will consider Gauss diagram formulas for R III moves. There are three sorts of R III moves: without any persistent crossings in the triangle, with exactly one persistent crossing, and with all three crossings in the triangle persistent. The first sort does not contribute at all.

Let an integer $0 < a < n$ be fixed. We consider in the 1-cocycle from the distinguished couple of persistent crossings *only* R III moves p of the global types $r(a, n, a)$ and $r(n, n, n)$, compare, e.g., Figs. 1.12 and 1.13 (where we give already the refined linking numbers).

Remark 1.2 *The edges of the tetrahedron correspond to the crossings and the 2-faces correspond to the triple crossings. If there is one 2-face corresponding to $r(a, n, a)$ or $r(n, n, n)$, then the remaining three edges have a common vertex. The homology class of one of these edges (i.e., the homological marking of the corresponding crossing) determines now the homology classes of the remaining two edges. Of course, the homology classes of three edges with a common vertex determine the homology classes of the remaining three edges.*

We call a tetrahedron *generic* if it contains one stratum of triple crossings with all markings in $\{0, n\}$ and there is at least one crossing

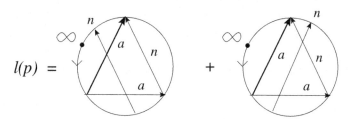

Figure 1.12: The refined linking number $l(p)$ for the type $r(a, n, a), \infty \in d^+$.

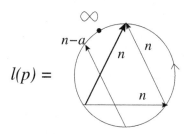

Figure 1.13: The refined linking number $l(p)$ for the type $r(n, n, n), \infty \in d^+$.

of marking a (there are then automatically two other crossings with markings in $\{a, n - a\}$ and there are no other triple crossings with all markings in $\{0, n\}$).

We call the tetrahedron *degenerate* if all six crossings have their markings in $\{0, n\}$, i.e., they are all persistent crossings.

Our strategy is the following: we solve first the generic tetrahedron equations by using essentially the weights $W_{2k}(p)$. We take then this solution and possibly complete it to a solution of the degenerate tetrahedron equations.

It turns out that for our 1-cocycle, we actually do not have to complete the solution in the degenerate case but just to check that it is still a solution without adding new strata or new weights.

Using Proposition 1.1, we could see our 1-cocycles from a dual point of view: we attach integer weights to the strata $\Sigma_{\text{tri}}^{(1)}$ in M_n which are completely determined by a single point in each smooth stratum $\Sigma_{\text{tri}}^{(1)}$. The weights have to fit together in the strata $\Sigma_{\text{quad}}^{(2)}$ and $\Sigma_{\text{trans-self}}^{(2)}$ in order to define an integer non-compact cycle of codimension 1 in the infinite dimensional space M_n. (The closure of this cycle in the closure $\bar{M}_n$ of M_n in the space of *all* knot diagrams in the solid torus with a fixed point ∞ would have its boundary in $\Sigma_{\text{trans-cusp}}^{(2)}$ and in $\Sigma_{\text{tri}-\infty}^{(2)}$, i.e., ∞ is in a crossing at the same moment as an R III move somewhere in the diagram happens, and $\Sigma_{\text{tan}}^{(1)}$ where two new crossings with a negative homology class appear.) Our 1-cocycles are then given by the weighted intersection numbers of smooth oriented generic loops in M_n with this cycle of codimension 1.

In the construction of our 1-cocycle from distinguished couples of persistent crossings, we glue together only strata $\Sigma_{\text{tri}}^{(1)}$ with exactly one persistent crossing with a single exception, the strata $r(n, n, n)$, which contributes just once in all the global tetrahedron equations!

But in the construction of our 1-cocycle from distinguished triples of persistent crossings, we always only glue together strata $\Sigma_{\text{tri}}^{(1)}$ with one persistent crossing with strata $\Sigma_{\text{tri}}^{(1)}$, where all three crossings are persistent.

The hardest part of future work will be to show that our 1-cochain from distinguished triples of persistent crossings is actually a solution of the degenerate global tetrahedron equations for all natural numbers $k > 1$.

Chapter 2

The 1-Cocycles from the Conway Polynomial

2.1 Generalities for R III Moves

The local types of Reidemeister moves for unoriented knots are shown in Fig. 2.1.

For oriented knots, there are eight local types of R III moves (compare Fig. 2.2), four local types of R II moves and four local types of R I moves. Different local types of Reidemeister moves come together in strata of $\Sigma^{(2)}$. Their relations will be extensively studied in Sections 3.2 and 6.6.

To each Reidemeister move of type III corresponds a diagram with a *triple crossing p*: three branches of the knot (the highest, middle and lowest with respect to the projection $pr : \mathbb{C}^* \times \mathbb{R} \to \mathbb{C}^*$) have a common point in the projection into the plane. A small perturbation of the triple crossing leads to an ordinary diagram with three crossings near $pr(p)$.

Definition 2.1 *We call the crossing between the highest and the lowest branch of the triple crossing p the* distinguished crossing *of p and we denote it by d (d stands for distinguished). The crossing between the highest branch and the middle branch is denoted by hm and that of the middle branch with the lowest is denoted by ml, compare Fig. 1.7. For better visualisation, we draw the crossing d always with a thicker arrow.*

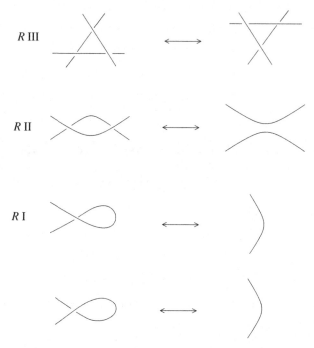

Figure 2.1:   The Reidemeister moves for unoriented knots.

Smoothing an ordinary crossing c of a diagram D with respect to the orientation splits the closure of D into two oriented and ordered circles. We call D_c^+ the component which goes from the undercross to the overcross at c and by D_c^- the remaining component, compare Fig. 1.4.

We associate to each generic crossing q in a diagram the integer $[q] = [D_q^+] \in \mathbb{Z} = H_1(V)$ for the above-mentioned identification. The integer $[q]$ is called the *homological marking* of q.

A *Gauss diagram* of a knot $K \subset V$ is an oriented circle with oriented chords which are decorated with their homological markings. The chords correspond to the crossings of the knot diagram and are always oriented from the undercross to the overcross (here we use the orientation of the $\mathbb{R}$-factor). Moreover, each chord (or arrow) has a sign, which corresponds to the writhe (or sign) of the crossing. The circle of a Gauss diagram in the plane is always equipped with the counterclockwise orientation.

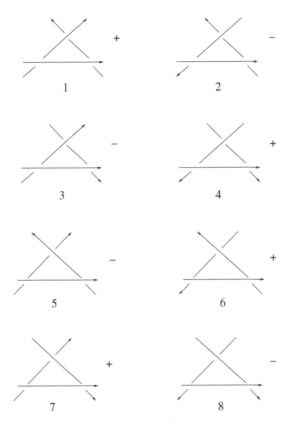

Figure 2.2: Local types of a triple crossing (with the local coorientations, which are used in Chapter 6).

A *Gauss diagram formula* of degree k is an expression assigned to the diagram of a knot $K \subset V$, which is of the following form:

$$\sum_{\text{configurations}} \text{function (writhes of the crossings)}$$

where the sum is taken over all possible choices of k (unordered) different crossings in the diagram such that the chords arising from these crossings in the diagram of K build a given sub-diagram with given markings but without fixing the signs on the arrows. The marked sub-diagrams are called *configurations*, compare, e.g., Refs. [15,40]. If the function is (as usual) the product of the writhes of the crossings in the configuration, then we will denote the sum shortly by the configuration itself. As usual, the *writhe* (or handedness, or sign) of a positive crossing (see, e.g., Fig. 1.4) is $w = +1$ and the writhe of a negative crossing is $w = -1$.

A Reidemeister III move corresponds to a triangle in the Gauss diagram. The *global type of a Reidemeister III move* is shown in Fig. 1.11, where $m + h$, $m + h - [K]$, m and h are the homological markings of the corresponding arrows. Here, $[K]$ is the homology class represented by the knot K. Using the homological markings for a given $n \neq 0$, we denote the corresponding strata in $\Sigma_{\mathrm{tri}}^{(1)}$ simply by $\Sigma_{([d],[hm],[ml])}^{(1)}$. Only for $n = 0$ we have in addition to indicate whether the arrow ml in the triangle goes to the left, denoted by l, or it goes to the right, denoted by r. We will consider Gauss diagram formulas for R III moves. The arrows in the configurations which are not arrows of the triangle are called the *weights* of the formulas. Moreover, the point ∞ is in one of the three open arcs in the circle for the triple crossing. Remember that for an n-crossing c, the point ∞ is always in c^+, and for a 0-crossing c, it is always in c^-.

Definition 2.2 *The coorientation for a Reidemeister III move is the direction from two intersection points of the corresponding three arrows to one intersection point and of no intersection point of the three arrows to three intersection points, compare Fig. 2.3. (We will see later in the cube equations for $\Sigma_{\mathrm{trans-self}}^{(2)}$ that the two coorientations for triple crossings fit together for the strata of $\Sigma_{\mathrm{tri}}^{(1)}$ which come together in $\Sigma_{\mathrm{trans-self}}^{(2)}$.) Evidently, our coorientation is completely determined by the corresponding planar curves and, therefore, we can draw just chords instead of arrows in Fig. 2.3. We call the side*

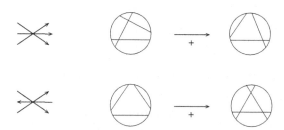

Figure 2.3:   The coorientation for Reidemeister III moves.

of the complement of $\Sigma^{(1)}$ in M_n into which points the coorientation the positive side *of $\Sigma^{(1)}$.*

The coorientation for Reidemeister II and Reidemeister I moves is the direction from no crossings to the diagram with two, respectively, one new crossing.

Each transverse intersection point p of an oriented generic arc in M with $\Sigma^{(1)}_{\text{tri}}$ has now an intersection index $+1$ or -1, called $sign(p)$, by comparing the orientation of the arc with the coorientation of $\Sigma^{(1)}$.

2.2 The Combinatorial 1-Cocycles by Using Distinguished Couples of Persistent Crossings

We will define four polynomial valued combinatorial 1-cocycles, which are called $\nabla R_{a,d^+}$, $\nabla R_{a,ml^-}$ and $\nabla R_{a,d^+,hm}$, $\nabla R_{a,ml^-,hm}$. All our polynomial valued 1-cocycles $\nabla R_{a,\ldots} \in H^1(M_n; \mathbb{Z}[z])$ for each $0 < a < n$ are of the form $\nabla R_{a,\ldots} = \sum_{k=0} R^{(2k)}_{a,\ldots} z^{2k}$, where $R^{(2k)}_{a,\ldots}$ is an integer 1-cocycle which is based on the formulas C_{2k} of Chmutov, Khoury and Rossi for each natural number k.

The 1-cocycles $R^{(2k)}_{a,\ldots}$ depend, in general, non-trivially on the integer parameter $0 < a < n$, and hence, they can be made polynomial-valued 1-cocycles of degree $n-2$ by taking the Lagrange interpolation polynomial with respect to the parameter a for each $k > 0$. Hence, the ∇R become two-variable polynomials.

Each component of M_n corresponds to a regular isotopy type of a knot K in the solid torus. Let us call this component M_K and let us consider those $K = T \cup nK'$, where K' is a regular isotopy class of a long knot in 3-space. Each of the loops $rot(K)$, $push(T)$ and $fh(K)$ can be realised in such a way that the maximal number of crossings of a knot in the loop is minimal among all loops homologous to it. This maximal number of crossings depends only on the knot type K. Consequently, the sum in the definition of $\nabla R_{a,\ldots}$ is finite for each K and, in particular, $\nabla R_{a,\ldots}$ can be calculated with polynomial complexity with respect to the number of crossings of K.

For the definition of $R^{(2k)}_{a,d^+}$ and $R^{(2k)}_{a,ml^-}$, we use only two global types of Reidemeister III moves: $r(a, n, a)$ and $r(n, n, n)$, compare Figs. 1.12 and 1.13. The weight $W_{2k}(p)$ was defined in Definition 1.6

and will be used only for the global type $r(a, n, a)$. The *point at infinity* enters also in the global type of the R III move: for the global type $r(a, n, a)$, it is necessarily in hm^+, hence it could be in d^+ or in ml^-. For the global type $r(n, n, n)$, it is necessarily in d^+ (and hence in hm^+ too). The sign of the Reidemeister III move p, denoted by $sign(p)$, was defined in Definition 2.2, compare Fig. 2.3.

Definition 2.3 *The weight $W_{2k-1}(hm)$ for a diagram with a triple crossing p is defined by the sum of all A_{2k} where the crossing hm in p is the first n-crossing in A_{2k}.*

Note that for the type $r(a, n, a)$, the crossings d and ml have the marking $a \neq n$ and hence do not contribute to A_{2k}. For the type $r(n, n, n)$, the crossings d and ml have the marking n. However, the point at infinity is necessarily in d^+. Consequently, if d or ml would contribute in a configuration of A_{2k}, then hm would not be the first n-crossing in the configuration, which is excluded.

Note also, that for the weight $W_{2k-1}(hm)$, the last 0-crossing is always automatically in the open arc hm^+.

Definition 2.4 *Let the triple crossing p be of type $r(a, n, a)$ with $\infty \in d^+$. Then the refined linking number $l(p)$ is defined as the sum of the signs of all n-crossings which cut the crossing ml of p (i.e., the corresponding arrows in the Gauss diagram intersect) and which have their foot in the open arc ml, compare Fig. 1.12.*

Let the triple crossing p be of type $r(n, n, n)$ with necessarily $\infty \in d^+$. Then the refined linking number $l(p)$ is defined now as the sum of the signs of all $(n - a)$-crossings (i.e., $[q] = n - a$) which cut the crossing ml of p (i.e., the corresponding arrows in the Gauss diagram intersect) and which have their foot in the open arc ml, compare Fig. 1.13 (there is only one possibility to cut the triangle for the $(n - a)$-crossings here because there are no negative loops in the diagrams).

Remember that $w(q)$ is the sign for an ordinary crossing q and that we consider the product of the signs of the $2k$ crossings for each $2k$-tuple of crossings in a diagram, which represent a configuration in A_{2k} in the Gauss diagram. We sum up over all such configurations.

We are now ready to define our first 1-cochain.

Definition 2.5 *Let γ be a generic oriented loop in M_n.*
The integer-valued 1-cochain $R^{(2k)}_{a,d^+}$ is defined for each natural numbers $n > 1$, $0 < a < n$ and $k > 0$ by

$$R^{(2k)}_{a,d^+}(\gamma) = \sum_{p=(r(a,n,a),\infty \in d^+) \in \gamma} \text{sign}(p)(W_{2k}(p)$$

$$+ (l(p) + w(hm) - 1)W_{2k-1}(hm)w(hm))$$

$$- \sum_{p=r(n,n,n) \in \gamma} \text{sign}(p)l(p)W_{2k-1}(hm)w(hm).$$

We want to apply the 1-cochains to canonical loops in M_n. In fact, it turns out that we can slightly weaken the condition of regular isotopy to *semi-regular isotopy*: in a R I move, a new crossing appears or disappears which has marking 0 or n. In a semi-regular isotopy, we allow R I moves with the marking 0, but we do not allow R I moves with the marking n, and we call the corresponding moduli space $M_n^{\text{semi-reg}}$. Note that for semi-regular isotopy of a long knot K, we have a finite type invariant $w_1(K)$ of order 1, namely, the algebraic sum of all crossings in K with marking 1.

Definition 2.6 *Let K' be an oriented framed long knot and let T be a string link which induces a cyclical permutation of its end points. We denote by $K = T \cup nK'$ the knot in M_n which is obtained by closing the parallel n-cable of K' (with respect to the framing and with the induced orientation) by the string link T to a knot in the solid torus. The loop push (T, nK') is defined by pushing T once through the parallel n-cable of K' in the solid torus in counterclockwise direction [16].*

Adding to $T \cup nK'$ a full twist in form of a positive n-curl, compare Fig. 2.4, we could then push $T \cup nK'$ once through the curl. This is an example for our pairing, where the long knot is just a positive curl with $w_1 = 0$. On the other hand, it is a nice representative in M_n of Gramain's loop $rot(K = T \cup nK')$, see Ref. [23], which is induced by the full rotation of V^3 around its core.

(Instead of classical knots, our invariants are of course also defined for arbitrary coherently oriented n-component string links T in $\mathbb{R}^3$, which induce a cyclic permutation of their endpoints, by evaluating the 1-cocycles, e.g., on $rot(T)$.)

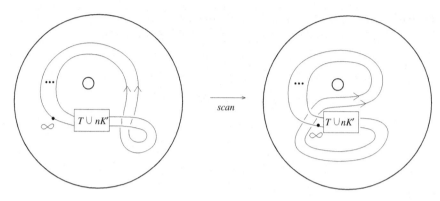

Figure 2.4:   Pushing $T \cup nK'$ through an n-curl.

One easily sees that the n-curl is semi-regular isotopic to the full twist Δ^2 (the generator of the centre of the braid group) in the braid group B_n. Hence, if T is a braid, then we could push also the n-curl through $T \cup nK'$. Geometrically, this corresponds to the loop which is generated by the full rotation of the companion solid torus $V_1^3 \subset V^3$ (from the cabling construction) around its core.

The following theorem is our principal result.

Theorem 2.1 $R_{a,d^+}^{(2k)}$ *is an integer 1-cocycle in* $M_n^{\text{semi-reg}}$ *for each* $n > 1$, *each* $0 < a < n$, *each* $k > 0$ *and which represents a non-trivial cohomology class. In particular, the corresponding pairings* $H_0(M_n^{\text{semi-reg}}) \times H_0(M_1^{\text{semi-reg}}) \to \mathbb{Z}$ *are non-trivial already for* $n = 2$ *and when* T *in* $T \cup 2K'$ *is just the standard generator* σ_1 *of the braid group* B_2.

We obtain another 1-cocycle by fixing now the point $\infty \in ml^-$ for the global type $r(a, n, a)$. But we have to change here the definitions of $l(p)$ as shown in Figs. 2.5 and 2.6, and we denote these refined linking numbers by $l_{ml^-}(p)$. Note that for the global type $r(n, n, n)$, the crossings d and ml can still not contribute to $W_{2k-1}(hm)$ because hm has to be the first n-crossing in the configurations.

$$l_{ml^-}(p) =$$ 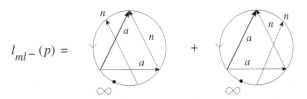 $$+$$

Figure 2.5: Refined linking number for the type $r(a, n, a)$ when $\infty \in ml^-$.

$$l_{ml^-}(p) =$$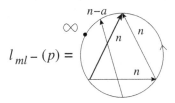

Figure 2.6: Refined linking number for the type $r(n, n, n)$ when $\infty \in ml^-$.

Definition 2.7 *Let γ be a generic oriented loop in M_n.*
The integer-valued 1-cochain $R^{(2k)}_{a,ml^-}$ is defined for each natural numbers $n > 1$, $0 < a < n$ and $k > 0$ by

$$R^{(2k)}_{a,ml^-}(\gamma) = \sum_{p=(r(a,n,a),\infty \in ml^-)\in\gamma} \text{sign}(p)$$

$$\times \left(W_{2k}(p) + (l_{ml^-}(p) + w(hm) - 1)W_{2k-1}(hm)w(hm)\right)$$

$$- \sum_{p=r(n,n,n)\in\gamma} \text{sign}(p)l_{ml^-}(p)W_{2k-1}(hm)w(hm).$$

We have the following theorem which is the analogue of Theorem 2.1.

Theorem 2.2 $R^{(2k)}_{a,ml^-}$ *is an integer 1-cocycle in $M_n^{\text{semi}-\text{reg}}$ for each $n > 1$, each $0 < a < n$, each $k > 0$ and which represents a non-trivial cohomology class.*

Let $\tilde{M}_n$ be the space of all knot diagrams in the solid torus and not only regular ones, i.e., in $\tilde{M}_n$ all R I moves are allowed as well. The following is the most simplest 1-cocycle from the Chmutov–Khoury–Rossi formulas and it is even well defined in $\tilde{M}_n$.

Definition 2.8 *Let γ be a generic oriented loop in $\tilde{M}_n$.*
The integer-valued 1-cochain $R^{(2k)}_{a,d^+,hm}$ is defined for each natural numbers $n > 1$, $0 < a < n$ and $k > 0$ by

$$R^{(2k)}_{a,d^+,hm}(\gamma) = \sum_{p=(r(a,n,a),\infty \in d^+) \in \gamma} \text{sign}(p) W_{2k-1}(hm) w(hm).$$

This means that we consider only those contributions to A_{2k}, where hm is the first n-crossing. In this case, we do not need the refined linking numbers neither the contributions from the global type $r(n, n, n)$.

Theorem 2.3 $R^{(2k)}_{a,d^+,hm}$ *is an integer 1-cocycle in $\tilde{M}_n$ for each $n > 1$, each $0 < a < n$, each $k > 0$ and which represents a non-trivial cohomology class.*

The definition of this 1-cocycle is the closest to the definition of the Conway polynomial as a 0-cocycle by Chmutov–Khoury–Rossi.

Let us prove that $R^{(2k)}_{a,d^+,hm}$ is well defined in $\tilde{M}_n$ if it is well defined in M_n. Evidently, a $R\,\mathrm{I}$ move does not change the weights and only $R\,\mathrm{I}$ moves with homological marking n could give a new contribution to $R^{(2k)}_{a,d^+,hm}$. There is only a single global type of triple crossings in the definition of the 1-cocycle. The $R\,\mathrm{I}$ move could lead to new triple crossings of type $r(a, n, a)$ only if the crossing from the $R\,\mathrm{I}$ move is the crossing hm in $r(a, n, a)$, compare Fig. 2.7.

But the crossing hm is isolated with respect to the other persistent crossings (it bounds just a small curl) and hence does not contribute to any A_{2k}. Consequently, the new triple crossing does not contribute to $R^{(2k)}_{a,d^+,hm}$ at all.

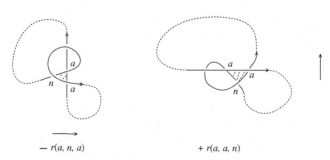

$$- \, r(a, n, a) \qquad\qquad + \, r(a, a, n)$$

Figure 2.7:   Sliding a small n-curl twice through a crossing of marking a.

We define the 1-cocycle $R^{(2k)}_{a,ml^-,hm}$ exactly in the same way as $R^{(2k)}_{a,d^+,hm}$, but we require that ∞ is now in ml^-.

Definition 2.9 *Let γ be a generic oriented loop in $\tilde{M}_n$.*
The integer-valued 1-cochain $R^{(2k)}_{a,ml^-,hm}$ is defined for each natural numbers $n > 1$, $0 < a < n$ and $k > 0$ by

$$R^{(2k)}_{a,ml^-,hm}(\gamma) = \sum_{p=(r(a,n,a),\infty \in ml^-) \in \gamma} \text{sign}(p) W_{2k-1}(hm) w(hm).$$

This cocycle also represents a non-trivial cohomology class.

Theorem 2.4 $R^{(2k)}_{a,ml^-,hm}$ *is an integer 1-cocycle in $\tilde{M}_n$ for each $n > 1$, each $0 < a < n$, each $k > 0$ and which represents a non-trivial cohomology class.*

Corollary 2.1 $R^{(2k)}_{a,d^+} + R^{(2k)}_{a,ml^-}$ *as well as* $R^{(2k)}_{a,d^+,hm} + R^{(2k)}_{a,ml^-,hm}$ *are non-trivial 1-cocycles in M_n for each natural numbers $n > 1$, $0 < a < n$ and $k > 0$, and which do not depend on the choice of the point at infinity (even if we have always to chose a point at infinity).*

Remember that for $k = 1$, we do not need the point at infinity at all because the homological markings determine in this case the arc which contains ∞. Consequently, the sum of the 1-cocycles, given in Definitions 2.5 and 2.7 for $k = 1$, is now a 1-cocycle which does *not* depend on the point at infinity (which is necessarily in hm^+ for the type $r(a,n,a)$ and in d^+ for the type $r(n,n,n)$ because it is always in c^+ for each crossing c of marking $[c] = n$).

It follows immediately from their definitions that $R^{(2)}_{a,d^+} + R^{(2)}_{a,ml^-}$ coincides with the 1-cocycle, which was called $R^{(2)}_a$, in Ref. [17]. This splitting of $R^{(2)}_a$ is non-trivial, however we haven't calculated enough examples in order to show that $R^{(2)}_{a,d^+}$ and $R^{(2)}_{a,ml^-}$ represent independent cohomology classes.

Evidently, $R^{(2k)}_{a,d^+} + R^{(2k)}_{a,d^+,hm}$ is just the 1-cocycle for which the weights $W_{2k}(p)$ are defined with the foot of the first n-crossings in the *closure* of the arc hm^+. This 1-cocycle will be generalised for distinguished triples of persistent crossings in Section 2.3.

A very interesting question is whether at least the 1-cocycle $R_{a,d^+,hm}^{(2k)}$ could be generalised for other quantum knot polynomials. Here is a first candidate:

Let C_3 be the Gauss diagram formula (in fact, its mirror image) for the Vassiliev invariant v_3 of long knots, which was found by Chmutov and Polyak in Ref. [12], as a particular case of the finite type invariants which can be extracted from the HOMFLYPT polynomial. C_3 has also the first–last foot property (as well as the identity, which is used for the proof of its invariance under R III moves and which holds only for real knots and not for virtual knots). But it turns out that it has even the following refined property.

Refined first–last foot property: going from ∞ along the circle, we come first to the foot of an n-crossing c and last to the foot of an 0-crossing. We consider now the *symbol* of C_3, i.e., all configurations in it which contain exactly three crossings. The foot of the remaining (middle) crossing q in the symbol is in c^- if and only if q is an n-crossing.

C_3 no longer has the preserving intersection property, but it is clear that it should be replaced by a more complicated property.

We define now the *weight* $W_2(hm)$ for a diagram with a triple crossing p of global type $r(a,n,a)$ by the sum of all those configurations in C_3 where the crossing hm in p is the first n-crossing in the configuration in C_3, and hence, the last 0-crossing has automatically its foot in hm^+.

Question 2.1 *Let γ be a generic oriented loop in $\tilde{M}_n$, $n > 1$.*
The integer-valued 1-cochain $R_{a,d^+,hm}^3$ is defined for each natural number $0 < a < n$ by

$$R_{a,d^+,hm}^3(\gamma) = \sum_{p=(r(a,n,a),\infty\in d^+)\in\gamma} \mathrm{sign}(p)W_2(hm)w(hm).$$

Is $R_{a,d^+,hm}^3$ also a non-trivial integer 1-cocycle in $\tilde{M}_n$?

Of course, one could then try to go on the same way with all the other Gauss diagram formulas for finite type invariants extracted from the HOMFLYPT polynomial in Ref. [12].

Remark 2.1 *A very important point is that for each knot, the Conway polynomial contains only a finite number of finite type invariants. This has as a consequence that the whole system of our* *invariants, which are based on 1-cocycles from the Conway polynomial, are calculable with polynomial complexity with respect to the crossing numbers of the diagrams, if we fix the number n of strands in the cabling construction. Let us call this property of the system of invariants polynomiality for fixed n. As is well known, the coloured Jones polynomial of the trefoil already contains an infinite number of finite type invariants. Hence, 1-cocycles from the coloured Jones polynomial would no longer have the "polynomiality for fixed n" property.*

However, Dror Bar-Natan and Roland van der Veen [3] have shown that the coloured Jones polynomial contains polynomials (probably the Rozansky polynomials up to normalisation) which can be calculated with polynomial complexity. It could perhaps be that these polynomials are related to ascending multi-component arrow diagrams A_k (exactly as the Conway polynomial is related to ascending one-component arrow diagrams), which we could perhaps put into our machinery too, in order to construct from them combinatorial 1-cocycles. They would again have the "polynomiality for fixed n" property.

2.3   A Combinatorial 1-Cocycle by Using Distinguished Triples of Persistent Crossings

In this sub-section we define a new 1-cocycle by using not only the first n-crossings but also at the same time the last 0-crossings in the configurations of A_{2k}. We call it $\bar{R}_a$, where the "bar" stands for completion.

This time we have to define the *symbol* $W_{2k}(p)$ of the weight of order $2k$ for two global types of triple crossings, namely, $r(a, n, a)$ with $\infty \in d^+$ and the type $l(a, 0, a)$ with $\infty \in ml^+$.

The weight $\bar{W}_{2k}(p)$ for the type $r(a, n, a)$ was already defined in Definition 1.7.

Definition 2.10 *The weight $\bar{W}_{2k}(p)$ for the global type $l(a, 0, a)$ is defined as the sum over all those arrow diagrams A_{2k} for which the first foot (of necessarily an n-crossing) is in the closure of hm^- and the last foot (of necessarily a 0-crossing) is in the closure of hm^- too.*

Note that for the weights $\bar{W}_{2k}(p)$, the crossing hm could be the first n-crossing, respectively, the last 0-crossing in a configuration of A_{2k}.

We have now to define the *degenerate* weights $\bar{W}_{2k-1}$.

Definition 2.11 *The weight $\bar{W}_{2k-1}(hm)$ for a diagram with a triple crossing p of global type $r(a,n,a)$ is defined by the sum of all A_{2k} where the crossing hm in p is the first n-crossing in A_{2k} (and the last 0-crossing in A_{2k} has necessarily its foot in hm^+ too).*

The weight $\bar{W}_{2k-1}(hm)$ for a diagram with a triple crossing p of global type $l(a,0,a)$ is defined by the sum of all A_{2k} where the crossing hm in p is the last 0-crossing in A_{2k} (and the first n-crossing in A_{2k} has necessarily its foot in hm^- too).

But we need also degenerate weights and refined linking numbers for those four degenerate global types of triple crossings, which mix 0-crossings and n-crossings. A triple crossing p corresponds to a generic point in the discriminant $\Sigma_{\mathrm{tri}}^{(1)}$ in M_n. The positive and negative sides of $\Sigma_{\mathrm{tri}}^{(1)}$ in M_n are defined in Fig. 2.3, and they depend only on the underlying planar curves. The degenerate weights depend now on two crossings in the triple crossing and moreover some of them depend on the side of the discriminant, which we consider in order to calculate them. We have, therefore, to specify the positive or negative side of the discriminant. The local types of triple crossings are shown in Fig. 2.2.

We distinguish two local sorts of triple crossings:

Braid-like, i.e., they can occur for braids. These are exactly the local types 1, 3, 4, 5, 7 and 8 (we can see them as the equator of the boundary of the cube with respect to the poles 2 and 6).

Non braid-like, i.e., they never occur for braids. These are exactly the local types 2 and 6.

It turns that we have to chose the positive side if and only if the triple crossing is braid-like. Moreover, we consider only open arcs for d^+ and for ml^+, i.e., the crossing d does not belong to d^+ and the crossing ml does not belong to ml^+.

Definition 2.12 *For a triple crossing p of global type $l(n,0,n)$, we have the following:*

The weight $\bar{W}_{2k-1}(d, ml^+)$ is defined by the sum of all A_{2k} where the crossing d in p is the first n-crossing in A_{2k} and the last 0-crossing in A_{2k} has its foot in ml^+ when we consider the positive side of p if and only if p is braid-like (otherwise, we consider of course the negative side).

The weight $\bar{W}_{2k-1}(hm, ml^+)$ is defined by the sum of all A_{2k} where the crossing hm in p is the last 0-crossing in A_{2k} and the first n-crossing in A_{2k} has its foot in ml^+ when we consider the positive side of p if and only if p is braid-like.

The weight $\bar{W}_{2k-1}(ml, d^+)$ is defined by the sum of all A_{2k} where the crossing ml in p is the first n-crossing in A_{2k} and the last 0-crossing in A_{2k} has its foot in d^+ when we consider the positive side of p if and only if p is braid-like.

Definition 2.13 *For a triple crossing p of global type $r(0, n, 0)$, we have the following:*

The weight $\bar{W}_{2k-1}(d, ml^-)$ is defined by the sum of all A_{2k} where the crossing d in p is the last 0-crossing in A_{2k} and the first n-crossing in A_{2k} has its foot in ml^- when we consider the positive side of p if and only if p is braid-like.

The weight $\bar{W}_{2k-1}(hm, ml^-)$ is defined by the sum of all A_{2k} where the crossing hm in p is the first n-crossing in A_{2k} and the last 0-crossing in A_{2k} has its foot in ml^- when we consider the positive side of p if and only if p is braid-like.

The weight $\bar{W}_{2k-1}(ml, d^-)$ is defined by the sum of all A_{2k} where the crossing ml in p is the last 0-crossing in A_{2k} and the first n-crossing in A_{2k} has its foot in d^- when we consider the positive side of p if and only if p is braid-like.

For the remaining two degenerate global types, the contributions do not depend on the chosen side of the discriminant.

Definition 2.14 *For a triple crossing p of global type $r(0, 0, n)$, we have the following:*

The weight $\bar{W}_{2k-1}(d, ml^+)$ is defined by the sum of all A_{2k} where the crossing d in p is the last 0-crossing in A_{2k} and the first n-crossing in A_{2k} has its foot in ml^+.

The weight $\bar{W}_{2k-1}(hm, ml^+)$ is defined by the sum of all A_{2k} where the crossing hm in p is the last 0-crossing in A_{2k} and the first n-crossing in A_{2k} has its foot in ml^+.

The weight $\bar{W}_{2k-1}(ml, d^-)$ is defined by the sum of all A_{2k} where the crossing ml in p is the first n-crossing in A_{2k} and the last 0-crossing in A_{2k} has its foot in d^-.

Definition 2.15 *For a triple crossing p of global type $l(n, n, 0)$, we have the following*:

The weight $\bar{W}_{2k-1}(d, ml^-)$ is defined by the sum of all A_{2k} where the crossing d in p is the first n-crossing in A_{2k} and the last 0-crossing in A_{2k} has its foot in ml^-.

The weight $\bar{W}_{2k-1}(hm, ml^-)$ is defined by the sum of all A_{2k} where the crossing hm in p is the first n-crossing in A_{2k} and the last 0-crossing in A_{2k} has its foot in ml^-.

The weight $\bar{W}_{2k-1}(ml, d^+)$ is defined by the sum of all A_{2k} where the crossing ml in p is the last 0-crossing in A_{2k} and the first n-crossing in A_{2k} has its foot in d^+.

Note that sometimes the weight is already determined just by the first crossing in the couple, for example, $W_1(d, ml^+) = W_1(d)$ for $r(0, 0, n)$ because no n-crossing with the foot in ml^- could contribute together with the 0-crossing d.

For the last two global types, the contributions of the weights do not depend on the side of the discriminant. Indeed, we have only to consider positive triple crossings, i.e., all three crossings in the triangle are positive (the remaining cases are treated only with the cube equations, see Chapter 3). Moreover, we have evidently only to study those configurations A_{2k}, which contain at least two of the three crossings of the triple crossing. Let us consider $\bar{W}_{2k-1}(d, ml^+)$ for the global type $r(0, 0, n)$. We show the positive and the negative side in Fig. 2.8. The 0-crossing d on the positive side cannot contribute together neither with ml (because ml is not in ml^+) nor with hm (because the foot of the 0-crossing hm would be closer to ∞ and hence d is not the last 0-crossing). The same is true on the negative side.

Let us consider $\bar{W}_{2k-1}(hm, ml^+)$ for the same global type. On the positive side, hm could contribute together with d in an A_{2k}. The foot of the first n-crossing in A_{2k} is then necessarily in the arc from ∞ to the head of hm. It follows then from the *preserving intersection property* (compare the Introduction) that on the negative side, hm and ml contribute together in C_{2k} (i.e., in another configuration A_{2k}, where just the crossing d is replaced by the crossing ml). But hm is

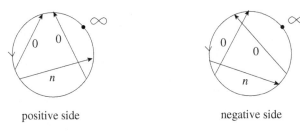

positive side   negative side

Figure 2.8: $\bar{W}_{2k-1}(d, ml^+)$ for the global type $r(0, 0, n)$ on both sides.

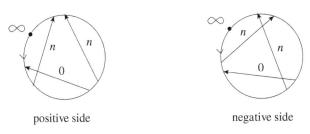

positive side   negative side

Figure 2.9: $\bar{W}_{2k-1}(d, ml^+)$ for the global type $l(n, n, 0)$ on both sides.

still the last 0-crossing. The n-crossing ml is allowed to contribute because it cannot be the *first* n-crossing.

Let us consider $\bar{W}_{2k-1}(ml, d^+)$ for the global type $l(n, n, 0)$. The positive and negative sides are shown in Fig. 2.9. The crossing ml is the last 0-crossing. But then it cannot contribute neither with d nor with hm because their heads are both in the interval from the foot of ml to ∞, no matter which side of the discriminant.

We have now to define the refined linking numbers but only for degenerate triple crossings (i.e., all three markings are in $\{0, n\}$). There will be only one type of linking number $l_a(p)$ for each $0 < a < n$. They will be given by combinations of configurations in figures. Remember that this means that the refined linking number is the sum of the signs of crossings a which represent the configuration in a Gauss diagram of the knot with a triple crossing in the solid torus. Consequently, for degenerate triple crossings, the linking numbers do not depend on the side of the discriminant. Remember also that there are no negative loops in diagrams and that the position of ∞ is not automatically determined for a crossing of marking a with $0 < a < n$.

Definition 2.16 *Let the triple crossing p be of type $l(n, 0, n)$ with necessarily $\infty \in ml^+$. Then, the refined linking number $l_a(p)$ is*

$$l_a(p) = $$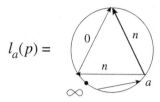

Figure 2.10:   Refined linking number for the type $l(n, 0, n)$.

$$l_a(p) = $$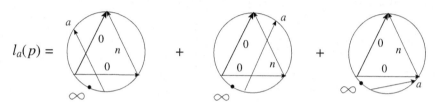

Figure 2.11:   Refined linking number for the type $r(0, n, 0)$.

$$l_a(p) = $$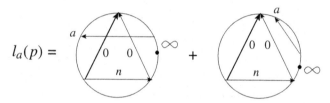

Figure 2.12:   Refined linking number for the type $r(0, 0, n)$.

defined as the sum of the signs of all a-crossings (i.e., $[q] = a$) shown in Fig. 2.10.

Definition 2.17 *Let the triple crossing p be of type $r(0, n, 0)$ with necessarily $\infty \in ml^-$. Then, the refined linking number $l_a(p)$ is defined as the sum of the signs of all a-crossings shown in Fig. 2.11.*

Definition 2.18 *Let the triple crossing p be of type $r(0, 0, n)$ with necessarily $\infty \in hm^-$. Then, the refined linking number $l_a(p)$ is defined as the sum of the signs of all a-crossings shown in Fig. 2.12.*

Definition 2.19 *Let the triple crossing p be of type $l(n, n, 0)$ with necessarily $\infty \in hm^+$. Then, the refined linking number $l_a(p)$ is defined as the sum of the signs of all a-crossings shown in Fig. 2.13.*

$$l_a(p) =$$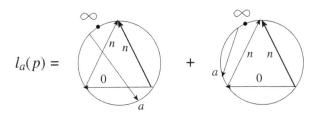

Figure 2.13:   Refined linking number for the type $l(n, n, 0)$.

Note that the refined linking numbers $l_a(p)$ are not determined by the foots of the a-crossings alone. Their definition is very non-symmetric, which reflects the fact that the weights of the symbol are *only* defined with respect to the crossing hm and not with respect to the crossing ml in the triple crossings. In fact, $l_a(p)$ is defined in the following way: the foot of the a-crossing is in the arc from the point ∞ and $\infty \in a^+$. The head of the a-crossing can slide over vertices of the triangle as far as there are no foots of two crossings in the vertex of the triangle.

We are now ready to define our new 1-cochain.

Definition 2.20 *Let γ be a generic oriented loop in M_n.*

The integer-valued 1-cochain $\bar{R}_a^{(2k)}$ is defined for each natural number $n > 1$, $0 < a < n$ and $k > 0$ by

$$\bar{R}_a^{(2k)}(\gamma) = \sum_{p=(r(a,n,a),\infty \in d^+) \in \gamma} \text{sign}(p)$$

$$\times (\bar{W}_{2k}(p) + 1/2(w(hm) - 1)\bar{W}_{2k-1}(hm)w(hm))$$

$$+ \sum_{p=(l(a,0,a),\infty \in ml^+) \in \gamma} \text{sign}(p)(\bar{W}_{2k}(p) + 1/2(w(hm) - 1)$$

$$\times \bar{W}_{2k-1}(hm)w(hm))$$

$$+ \sum_{p=(l(n,n,0),\infty \in hm^+) \in \gamma} \text{sign}(p)l_a(p)(\bar{W}_{2k-1}(d, ml^-)w(d)$$

$$- \bar{W}_{2k-1}(hm, ml^-)w(hm) - \bar{W}_{2k-1}(ml, d^+)w(ml))$$

$$+ \sum_{p=(l(n,0,n),\infty \in ml^+) \in \gamma} \text{sign}(p)l_a(p)(\bar{W}_{2k-1}(d, ml^+)w(d)$$

$$- \bar{W}_{2k-1}(hm, ml^+)w(hm) - \bar{W}_{2k-1}(ml, d^+)w(ml))$$

$$- \sum_{p=(r(0,0,n),\infty \in hm^-) \in \gamma} \text{sign}(p)l_a(p)(\bar{W}_{2k-1}(d, ml^+)w(d)$$

$$- \bar{W}_{2k-1}(hm, ml^+)w(hm) - \bar{W}_{2k-1}(ml, d^-)w(ml))$$

$$- \sum_{p=(r(0,n,0),\infty \in ml^-) \in \gamma} \text{sign}(p)l_a(p)(\bar{W}_{2k-1}(d, ml^-)w(d)$$

$$- \bar{W}_{2k-1}(hm, ml^-)w(hm) - \bar{W}_{2k-1}(ml, d^-)w(ml)).$$

We can now formulate our next important result.

Theorem 2.5 $\bar{R}_a^{(2k)}$ *is an integer 1-cocycle in* M_n *for each* $n > 1$, *each* $0 < a < n$, $k = 1$ *and which represents a non-trivial cohomology class.*

Note that $\bar{R}_a^{(2k)}$ breaks the symmetry too: the symbol $W_{2k}(p)$ is defined only with respect to the crossing hm in the triple crossings p and never with respect to the crossing ml.

The following is our main conjecture.

Conjecture 2.1 *The 1-cochain* $\bar{R}_a^{(2k)}$ *is an integer 1-cocycle in* M_n *for each* $n > 1$, *each* $0 < a < n$ *and each* $k > 0$.

Theorem 2.5 shows that the conjecture is true for $k = 1$ (and of course it is true for $k = 0$, compare Ref. [16]). The example in Chapter 4 shows then that $\bar{R}_a^{(2k)}$ would represent a non-trivial cohomology class in general.

It seems to be possible to prove the conjecture for non-degenerate tetrahedrons. The really hard parts are the degenerate tetrahedrons (i.e., all six crossings are persistent crossings). It seems that we would need an analogue of the *preserving intersection property* for tetrahedrons which generalizes that for triangles.

Finally, let us mention that we could replace $r(a, n, a), \infty \in d^+$ by $r(a, n, a), \infty \in ml^-$ and $l(a, 0, a), \infty \in ml^+$ by $l(a, 0, a), \infty \in d^-$ and try to construct the corresponding 1-cocycle analogue to $\bar{R}_a^{(2)}$. But we haven't carried this out.

2.4   An Essential Refinement of the Combinatorial 1-Cocycles

In the tetrahedron equation, only four branches of the knot occur, and in the commutation relations and in the cube equations, even only three branches occur. This enables us to make a connection between the a-crossings, which serve as a parameter.

We consider the knot in the solid torus as an n-component string link T in 3-space, which is closed by a cyclic permutation of its end points in the two discs at infinity. Let us colour the n-components of T by C_1 up to C_n, where we start from the distinguished point ∞ with the component C_1.

Definition 2.21 *An a-crossing is of type C_i, for $i \in \{1, 2, \ldots, n\}$, if the* foot *of the a-crossing is in C_i.*

An $(n - a)$-crossing is of type C_i in the refined linking number $l(p)$ for the global type $r(n, n, n)$ if the head *of the $(n - a)$-crossing is in C_i.*

An a-crossing is of type C_i in the refined linking numbers $l_a(p)$ for $\bar{R}_a^{(2k)}$ if the foot *of the a-crossing is in C_i.*

We denote all these a-crossings simply by a_i, respectively, $(n-a)_i$.

This allows us to define splittings of all our combinatorial 1-cocycles into sums of 1-cocycles.

Definition 2.22 *Let γ be a generic oriented loop in M_n.*

The integer-valued 1-cochain $R_{a_i,d^+}^{(2k)}$ is defined for each natural numbers $n > 1$, $0 < a < n$, $1 \leq i \leq n$ and $k > 0$ by

$$R_{a_i,d^+}^{(2k)}(\gamma) = \sum_{p=(r(a_i,n,a_i),\infty \in d^+) \in \gamma} sign(p)(W_{2k}(p)$$

$$+ (l(p) + w(hm) - 1)W_{2k-1}(hm)w(hm))$$

$$- \sum_{p=r(n,n,n) \in \gamma} sign(p)l(p)W_{2k-1}(hm)w(hm).$$

The refined linking number $l(p)$ for the type $r(n, n, n)$ is here of course defined by using only $(n - a)_i$-crossings.

Exactly in the same way we refine all our other 1-cocycles to 1-cocycles $R_{a_i,d^+,hm}^{(2k)}$, $R_{a_i,ml^-}^{(2k)}$, $R_{a_i,ml^-,hm}^{(2k)}$ and $\bar{R}_{a_i}^{(2k)}$.

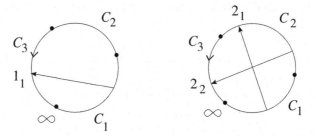

Figure 2.14: The possible a_i-crossings for $a = 1$ and $a = 2$ if $n = 3$.

The following theorem is our most important result.

Theorem 2.6 $R^{(2k)}_{a_i,d^+}$, $R^{(2k)}_{a_i,d^+,hm}$, $R^{(2k)}_{a_i,ml^-}$, $R^{(2k)}_{a_i,ml^-,hm}$ and $\bar{R}^{(2)}_{a_i}$ are *1-cocycles in M_n for each natural numbers $n > 1$, $0 < a < n$, $1 \le i \le n$ and $k > 0$.*

We have of course the splittings similar to $R^{(2k)}_{a,d^+} = \sum_{i=1}^{i=a} R^{(2k)}_{a_i,d^+}$ for all of our combinatorial 1-cocycles.

One easily sees that for each fixed $0 < a < n$ in the case of $R^{(2k)}_{a,d^+}$, there are exactly a different crossings a_i possible. We show as an example the case $n = 3$ for $a = 1$ and for $a = 2$ in Fig. 2.14. The dots are the intersections of the knot with the disc at infinity, and there is a distinguished dot ∞. (Remember that for each a-crossing q in $R^{(2k)}_{a,d^+}$, we need by definition $\infty \in q^+$.)

Let us mention that it turns out that there is no splitting possible by using the persistent crossings.

We haven't calculated any example by hand because there isn't yet any splitting for $n = 2$ and hence $a = 1$.

It turns out that the same refinement applies also to our former 1-cocycles from Ref. [16].

Let $R_{(n,b)}$ be the 1-cocycle for the open 1-simplex (n,b)), $0 < b < n$, which was defined in Chapter 2 of Ref. [16].

Definition 2.23 $R_{(n,b_j),d^+}$ *is the 1-cochain $R_{(n,b)}$, where we consider only those crossings of marking b with the head on the component C_j and with $\infty \in b^+$.*

We illustrate the definition in Fig. 2.15.

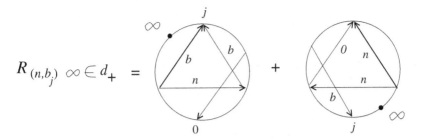

$$R_{(n,b_j)} \ \infty \in d_+ \ = \ \quad + \quad$$

Figure 2.15: The 1-cochain $R_{(n,b_j),d^+}$.

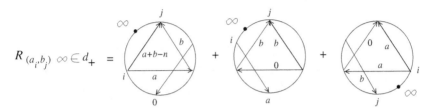

$$R_{(a_i,b_j)} \ \infty \in d_+ \ = \ \quad + \quad + \quad$$

Figure 2.16: The 1-cochain $R_{(a_i,b_j),d^+}$.

Let $R_{(a,b)}$ be the 1-cocycle for the open 2-simplex $(a,b))$, $0 < a < n$, $0 < b < n$, $a + b > n$, which was defined in Chapter 2 of Ref. [16].

Definition 2.24 $R_{(a_i,b_j),d^+}$ *is the 1-cochain* $R_{(a,b)}$, *where we consider only those crossings of marking* a *with the foot on the component* C_i, *we consider only those crossings* b *with the head on the component* C_j, *and with* $\infty \in a^+$ *and* $\infty \in b^+$.

We illustrate the definition in Fig. 2.16. Let us mention that for $a = b$, we consider the colours as in Fig. 2.16, i.e., we are not allowed to colour the foot and the head simultaneously for a crossing.

Proposition 2.1 $R_{(n,b_j),d^+}$ *and* $R_{(a_i,b_j),d^+}$ *are 1-cocycles in* $\tilde{M}_n$ *for all* $n > 0$, $0 < b < n$, $0 < a < n$, $a + b > n$, $1 \leq i, j \leq n$.

Note that $i = j$ is allowed and that $R_{(n,b_j),d^+}$ and $R_{(a_i,b_j),d^+}$ define of course splittings of $R_{(n,b)}$ and correspondingly of $R_{(a,b)}$ into sums of 1-cocycles.

Of course, $R_{(a,n)}$ from Chapter 2 in Ref. [16] can be refined in the same way by considering only those a-crossings with the foot on C_i. We haven't checked the refinement for the remaining 1-cocycles from Chapter 2 in Ref. [16].

2.5   A Refined 1-Cocycle for Cables of Long Knots Which Uses All Crossings

As before, we consider the knot in the solid torus as an n-component string link T in 3-space, which is closed by a cyclic permutation of its end points in the two discs at infinity. Let us colour the n-components of T by 1 up to n, where we start from the distinguished point ∞ with the component 1. But we allow now $n = 1$ too.

Instead of using only the persistent crossings for defining weights for R III moves, we use now *all* crossings of T.

Definition 2.25 *Let $i, j, k \in \{1, 2, \ldots, n\}$ be fixed integers, not necessarily all distinct, but $i \neq k$. We define the 1-cochain $R^{(2)}_{i,j,k}$ in Fig. 2.17.*

We use here our usual notation conventions: we sum up over all signed with sign(p) R III moves of the given types in the loop. To each of the moves we associate the sum of the signs of the crossings which correspond to the arrow, which is not in the triangle. These crossings cut always the crossing hm in the move in a configuration C_2. The sign of the crossing hm does not enter. The letters i, j, k at the foot of an arrow mean that the corresponding crossing has its foot (the undercross) on the corresponding component of T.

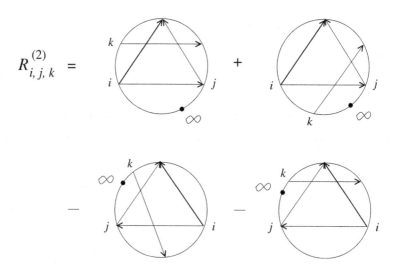

Figure 2.17:   The 1-cochain $R_{i,j,k}$ which uses all crossings.

If $i = k$, then we deform the triple crossing first into a regular diagram, no matter on the positive or negative side of the move, and we count also the contributions of d or ml if they form a configuration C_2 with the crossing hm (which happens only for the first two configurations in Fig. 2.17, compare the Introduction).

Let us call the arrow, which is not in the triangle, a k-crossing. We can now formulate our next theorem.

Theorem 2.7 $R_{i,j,k}^{(2)}$ *is a 1-cocycle in M_n for each natural numbers $n > 0$, $1 \le i, j, k \le n$. It represents a non-trivial cohomology class. If $i \ne j$, then it is a 1-cocycle even in $\tilde{M}_n$ (i.e., for all isotopies and not only regular isotopies).*

Remark 2.2 *The couple of markings $\{i, j\}$ determines always the homological marking of the crossing ml (compare the previous subsection), but for all other crossings, the homological marking is now arbitrary.*

If i, j, k are pairwise distinct, then they define a cyclical order on the circle in the Gauss diagram of the knot. Note that this cyclical order for the first configuration in the formula is different from the cyclical order for the three remaining configurations. It would be very interesting to study how the knot invariants depend on this cyclical order under changing the knot to its mirror image or to its inverse!

We show that $R_{i,j,k}^{(2)}$ represents a non-trivial cohomology class by a very simple example, namely, the loop $rot(K) \in M_1$ for the positive trefoil K just as a long knot, i.e., $n = 1$ and hence, $i = j = k = 1$ too.

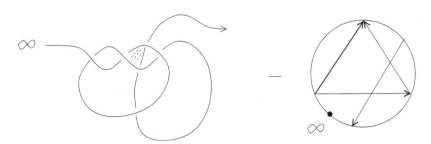

Figure 2.18: The calculation of $R_{1,1,1}^{(2)}(rot(K))$ for the positive trefoil.

We show the only non-trivial contribution in Fig. 2.18. It follows that $R_{1,1,1}^{(2)}(rot(K)) = -1$. (The contribution to C_2 comes from ml on the positive side or from d on the negative side of the move because $i = k = 1$ here.)

It seems to be unlikely that $R_{i,j,k}^{(2)}$ is a finite type 1-cocycle for knots in the solid torus in the sense of Vassiliev, see Refs. [46,47], because of the markings. If we switch the crossing ml, then i and j would interchange. But if we switch a k-crossing, then we loose all information because its head could be on a arbitrary component of T.

It would be very interesting to generalise $R_{i,j,k}^{(2)}$ to a 1-cocycle $R_{i,j,k}^{(2m)}$ for arbitrary degree $2m$. We have to replace C_2 by C_{2m} and as a first try, we define the corresponding weights in the following way:

Let us consider the generic case that i, j, k are pairwise distinct. A configuration in C_{2m} enters the weight if it contains the crossing hm and a k-crossing. Moreover, when we travel from ∞ along the circle, then we meet first the foot of hm and we meet last the foot of the k-crossing in C_{2m}.

We haven't studied the following natural question: Is the resulting 1-cochain already a 1-cocycle?

2.6   Potential Applications

2.6.1   The Conway polynomial as a combinatorial 1-cocycle

Let K be a long framed knot, let $\nabla_K(z)$ be its Conway polynomial and let $w_1(K)$ be the algebraic number of crossings of homological marking 1. We take $n = 2$ and hence $a = 1$. We consider the knot $\sigma_1 \cup 2K$ in the solid torus and the loop $push(\sigma_1)$ in $M_2^{\text{semi-reg}}$, which consists of pushing σ_1 twice through the 2-cable $2K$.

Conjecture 2.2 *For each long framed knot K, we have*

$$\nabla R_{a,d^+}(\text{push}(\sigma_1)) + \nabla R_{a,d^+,hm}(\text{push}(\sigma_1))$$

$$= \nabla R_{a,ml^-}(\text{push}(\sigma_1)) + \nabla R_{a,ml^-,hm}(\text{push}(\sigma_1)) = w_1(K)\nabla_K^2(z).$$

Moreover, $\nabla R_{a,d^+,hm}(\text{push}(\sigma_1)) = \nabla_K(z) - 1$.

We have checked the conjecture for three knots:
The trivial knot K with $w_1(K) = 1$.

$\nabla_K(z) = 1$
$\nabla R_{a,d^+}(\text{push}(\sigma_1)) = \nabla R_{a,ml^-}(\text{push}(\sigma_1)) = 1$
$\nabla R_{a,d^+,hm}(\text{push}(\sigma_1)) = \nabla R_{a,ml^-,hm}(\text{push}(\sigma_1)) = 0.$

The positive trefoil K with $w_1(K) = 1$.

$\nabla_K(z) = 1 + z^2$
$\nabla R_{a,d^+}(\text{push}(\sigma_1)) = 1 + z^2 + z^4$
$\nabla R_{a,d^+,hm}(\text{push}(\sigma_1)) = z^2$
$\nabla R_{a,ml^-}(\text{push}(\sigma_1)) = 1 + z^2$
$\nabla R_{a,ml^-,hm}(\text{push}(\sigma_1)) = z^2 + z^4.$

The positive torus knot K of type $(5,2)$ with $w_1(K) = 2$.

$\nabla_K(z) = 1 + 3z^2 + z^4$
$\nabla R_{a,d^+}(\text{push}(\sigma_1)) = 2 + 9z^2 + 21z^4 + 12z^6 + 2z^8$
$\nabla R_{a,d^+,hm}(\text{push}(\sigma_1)) = 3z^2 + z^4$
$\nabla R_{a,ml^-}(\text{push}(\sigma_1)) = 2 + 9z^2 + 12z^4 + 6z^6 + z^8$
$\nabla R_{a,ml^-,hm}(\text{push}(\sigma_1)) = 3z^2 + 10z^4 + 6z^6 + z^8.$

The configurations in C_{2k} for $k > 2$ which appear here in the calculations are almost all of the form of products of configurations of lower order (i.e., connected sums of configurations of lower order as in the last line of Fig. 1.9).

It follows that Conjecture 2.2 is true for these knots. However, it seems that the splitting does not give new knot invariants because in the above examples, we have also

$$\nabla R_{a,ml^-,hm}(\text{push}(\sigma_1)) = (\nabla_K(z) - 1)^2 + \nabla_K(z) - 1.$$

But of course, e.g., $\nabla R_{a,d^+,hm}(push(\sigma_1)) = \nabla_K(z) - 1$, distinguishes now loops up to homology in the moduli space. For example, if we change the orientation of the loop, then its value changes the sign. *So we can see the 1-cocycle $\nabla R_{a,d^+,hm}$ as an analogue of the Conway polynomial for loops of knots.*

Let K be the positive trefoil with $n = 2$, $a = 1$, $T = \sigma_1$ and let rot be the regular Gramain's loop as shown in Fig. 2.19, where we replace of course K by its twisted 2-cable $T \cup 2K$. We have calculated that

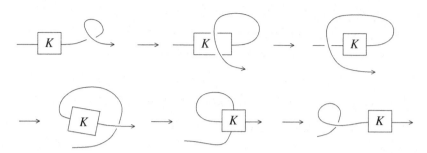

Figure 2.19:   A representative of Gramain's loop with a curl with a 1-crossing.

$$\nabla R_{a,d^+,hm}(rot(T \cup 2K)) = -z^2 - z^4 \text{ and}$$
$$\nabla R_{a,ml^-,hm}(rot(T \cup 2K)) = -z^2 - z^4.$$

The calculation by hand of $\nabla R_{a,d^+}(rot(T \cup 2K))$, which is of higher complexity than $\nabla R_{a,d^+,hm}(rot(T \cup nK))$, was already rather laborious (see Chapter 4). The calculation by hand of, e.g., $\nabla R_{a,d^+}(fh(T \cup 2K))$ is out of reach.

We would need much more examples in order to understand the relations of these 1-cocycles and their possible relations with the Conway polynomial of K.

2.6.2   Rational functions from 1-cocycles and torus knots

The most important loop in M_n is the Fox–Hatcher loop, compare Ref. [16] and references therein. It would also be very interesting to calculate these 1-cocycles on the Fox–Hatcher loop $fh(T \cup nK)$, compare the Introduction.

Because the 1-cocycles $\nabla R_{a,d^+,hm}$ and $\nabla R_{a,ml^-,hm}$ are well defined in $\tilde{M}_n$, we do not even need to approximate the Fox–Hatcher loop by a loop in M_n, compare Ref. [16] and references therein for the discussion of such approximations.

Let us recall the construction of the Fox–Hatcher loop for long knots. We go on K from ∞ to the first crossing. If we arrive at an undercross, then we move the branch of the overcross over the rest of the knot up to the end of K. If we arrive at an overcross, then we move the branch of the undercross under the rest of the knot up to its end. We continue the process up to the moment when we obtain a diagram which is isotopic to our initial diagram of K, compare

Ref. [25]. We can of course consider the analogue loop for cables of long framed knots $T \cup nK$ in $\tilde{M}_n$ by moving bunches of strands of nK without moving the string link T which was used to close the parallel n-cable to a knot in V^3.

As mentioned in the Introduction, Hatcher [25] has proven that for a long knot K, the homology classes $[rot(K)]$ and $[fh(K)]$ are linearly dependent on $H_1(\tilde{M}_1; \mathbb{Q})$ if and only if K is a torus knot.

In Ref. [16], we have proven the following lemma.

Lemma 2.1 *If K is a non-trivial torus knot, then the non-trivial classes $[rot(T \cup nK)]$ and $[fh(T \cup nK)]$ are linearly dependent on $H_1(\tilde{M}_n; \mathbb{Q})$ for each $n > 0$ and each closure to a knot in the solid torus with a string link T.*

This lemma implies immediately the following result.

If K is a non-trivial torus knot, then the rational function

$$\nabla R_{a,d^+,hm}(fh(T \cup nK)) / \nabla R_{a,d^+,hm}(rot(T \cup nK))$$

is in fact a rational number for each $n > 1$, each $0 < a < n$ and each string link T.

It would be very interesting to calculate this rational function for the figure eight knot. Is it still a constant? (We do not see any reason for this because the two loops are independent.) The same is of course also true for the rational function $\nabla R_{a,ml^-,hm}(fh(T \cup nK)) / \nabla R_{a,ml^-,hm}(rot(T \cup nK))$.

The analogue of this result for the 1-cocycle $\nabla R_{a,d^+}$ is a bit more complicated.

The absolute term of the Conway polynomial of a knot is always equal to 1. This is not the case for our 1-cocycles, but let us consider the 1-cocycles without the absolute term, denoted correspondingly by $\tilde{\nabla} R_{a,\ldots} = \sum_{k=1} R_{a,\ldots}^{(2k)} z^{2k}$. Let $\nabla_{T \cup nK}(z)$ denote the Conway polynomial of the associated long knot in the n-fold cyclic covering of the solid torus, compare the Introduction. If T is just a cyclic permutation braid, then of course $\nabla_{T \cup nK}(z) = \nabla_K(z)$, where $\nabla_K(z)$ is the usual Conway polynomial of the knot K in 3-space. If T is, e.g., itself an n-cable of a long knot K', which is twisted by a cyclic permutation braid, then $\nabla_{T \cup nK}(z) = \nabla'_K(z) \nabla_K(z)$. Let us denote $\nabla_{T \cup nK}(z) - 1$ shortly by $\tilde{\nabla}_{T \cup nK}(z)$.

The Fox–Hatcher loop is not a regular isotopy. But of course, by using Whitney tricks, it can be approximated by a regular isotopy, see, e.g., Ref. [15]. This approximation is not unique, but we show that for each component of M_n, the kernel of the natural inclusion $in\colon H_1(M_n) \to H_1(\tilde{M}_n)$ is generated by sliding small curls of marking 0 and of marking n along the whole knot.

Proposition 2.2 *Let in: $M_n \to \tilde{M}_n$ be the inclusion. Then, the kernel of $in_* : H_1(M_n) \to H_1(\tilde{M}_n)$ for each component of M_n is generated by sliding a small curl with marking 0 and sliding a small curl with marking n along the whole knot.*

Proof. Let $\gamma(s) \subset M_n$, $s \in [0,1]$, be a generic loop with base point $*$ which contracts in $\tilde{M}_n$. We try to contract it in M_n to $*$. Let γ_t, $t \in [0,1]$, be a generic homotopy of $\gamma_0 = \gamma$ to $\gamma_1 = *$ in $\tilde{M}_n$.

We have only to study the following events in the 2-parameter family $\gamma(s)_t$, $s, t \in [0,1]$, (compare Ref. [16]):

(1) γ_t becomes tangential to $\Sigma_{\text{cusp}}^{(1)} \subset M_n$. In case (a), a cusp is born and dies immediately after and in case (b), a cusp dies and is immediately reborn.

(2) γ_t passes through the transverse intersection of a stratum $\Sigma_{\text{cusp}}^{(1)}$ with another stratum of $\Sigma^{(1)}$.

(3) γ_t passes through $\Sigma_{\text{trans-cusp}}^{(2)}$.

(4) γ_t passes through $\Sigma_{\text{cusp-deg}}^{(2)}$. $\square$

In (1a), we replace the birth and the following death of a cusp in γ_{t_0} by a Whitney trick and its inverse and we continue the homotopy $\gamma_t \subset M_n$, $t > t_0$. The effect are just two additional small curls on the diagrams for a small arc s in the loops γ_t, $t > t_0$. We keep the small curls in the rest of the homotopy.

If γ_t touches $\Sigma_{\text{cusp}}^{(1)}$ from the other side, i.e., in (1b), then we simply keep the corresponding small curl on the arc s.

In (2), there is no problem at all because we can make Whitney tricks simultaneously at different places of a diagram.

In (3), a branch moves over or under a cusp. But we can perform a Whitney trick which creates or eliminates the cusp simultaneously with moving the brunch over or under the Whitney trick.

In (4), a Reidemeister II move is replaced by the birth or the death of two cusps with different writhe and different Whitney index.

We can replace this by a single Whitney trick, which is followed by sliding one of the two curls once over the other in order to obtain the diagram which allows the Reidemeister II move.

We end up with a diagram $*'$ which is regularly isotopic to $*$, but which differs from $*$ by lots of small curls on it. But because $*'$ and $*$ are regularly isotopic, the sum of the writhe's of the small curls as well as the sum of their Whitney indices vanishes. This implies that we can eliminate them two by two with Whitney tricks. This approximation of the homotopy γ_t by a homotopy in M_n is unique up to adding loops in M_n which consist of sliding small curls through small curls and sliding small curls all along the knot. These loops are evidently contractible in $\tilde{M}_n$, but our method does not show that they are contractible in M_n because we are not allowed to contract small curls (as in (1b)). □

Let U denote a diagram of an unknot which consists of small curls on a small arc of the knot K. In fact, one can prove that, e.g., Gramain's loop $rot(U)$, which is obtained by exchanging two positive curls with the same negative Whitney index, is not contractible in M_n by using the techniques of *trace graphs* developed in Ref. [19], compare also Ref. [16]. The trace graph of a loop in M_n is an oriented singular link in the thickened torus. All its singularities are ordinary triple points. The parity of the number of non-contractible components of any resolution of the trace graph in the thickened torus is an invariant of the homotopy class of the loop in M_n, compare Ref. [19]. The trace graph of the constant loop is just the standard closure of the trivial n-braid, where n is the number of crossings of the knot, hence $n = 2$ in our case of U. On the other hand, one easily sees that any resolution of the trace graph of $rot(U)$ has only one component because the two crossings are interchanged by the monodromy. This implies that the kernel of $in_* : H_1(M_n) \to H_1(\tilde{M}_n)$ is never trivial.

Proposition 2.3 *The value of the 1-cocycle* $\tilde{\nabla}R_{a,d^+}$ *is an integer multiple of* $\tilde{\nabla}_{T \cup nK}(z)$ *on each element in the kernel of* in_*: $H_1(M_n) \to H_1(\tilde{M}_n)$.

Proof. Sliding a small curl of marking 0 does not lead to any R III moves which contribute to $\tilde{\nabla}R_{a,d^+}$. We have only to consider the sliding of a curl of marking n. If a crossing is not of marking a, then sliding the curl over or under the crossing does not contribute neither. Let us consider a crossing of marking a. By using Whitney tricks and Reidemeister II moves, we can easily show that it is sufficient to consider a positive crossing and a positive curl. But the curl slides exactly one time over and one time under the crossing, as shown in Fig. 2.7. The first R III move here is of type $-r(a, n, a)$ and the second is of type $+r(a, a, n)$. As already mentioned, the crossing hm in $r(a, n, a)$ is isolated with respect to the remaining persistent crossings and hence, $W_{2k-1}(hm) = 0$. But *all* configurations in A_{2k} have all foots in the open arc hm^+ because the arc hm^- is almost empty (it contains just the foot or the head of another a-crossing). Consequently, all A_{2k} contribute now and $W_{2k}(p) = \tilde{\nabla}_{T \cup nK}(z)$, according to the Chmutov–Khoury–Rossi formula.

Let us consider the case that the curl of marking n slides over or under an n-crossing. If the crossing of the curl is the crossing hm in $r(n, n, n)$, then again it is isolated because the crossings d and ml cannot contribute to $W_{2k-1}(hm)$ (the crossing hm has to be the first n-crossing). The n-crossing from the curl cannot be the crossing d in $r(n, n, n)$, but it can be the crossing ml. In this case, $W_{2k-1}(hm)$ could be non-trivial because hm^- is not necessarily empty. However, the generalised linking number $l_{n-a}(p)$ is now 0 because the $n - a$-crossings have to cut ml and this is not possible because ml^- is empty.

We have proven that sliding small curls along the knot can only contribute integer multiples of $\tilde{\nabla}_{T \cup nK}(z)$ to $\tilde{\nabla}R_{a,d^+}$. $\qquad \square$

It follows from Lemma 2.1, Propositions 2.2 and 2.3 that if K is a non-trivial torus knot, then for each $n > 1$, each $0 < a < n$ and each string link T there exist integers m_1 and m_2 such that

$$(\tilde{\nabla}R_{a,d^+}(fh(T \cup nK)) + m_1 \tilde{\nabla}_{T \cup nK}(z))/(\tilde{\nabla}R_{a,d^+}(rot(T \cup nK))$$
$$+ m_2 \tilde{\nabla}_{T \cup nK}(z))$$

is in fact only a rational number!

We have of course the same results for $\tilde{\nabla}R_{a,ml^-}$. Moreover, we could consider the same quotients for our refined 1-cocycles (compare Section 2.4), as well as for $R_{i,j,k}^{(2)}$ (compare Section 2.5).

2.6.3 Candidates for distinguishing the knot orientation

A important point is that a'll our 1-cocycles in this chapter are a priory sensitive to the orientation of the knot.

Let $flip$ denote the rotation of the 3-space by π around the y-axis (i.e., it interchanges the two ends of a long knot) followed by an abstract orientation reversing of the knot. It is easy to see that a long knot K is isotopic to the long knot $-flip(K)$ as long knots if and only if the closure of K is an invertible knot in the 3-sphere, see, e.g., Ref. [15].

Consequently, if K is an invertible knot then, e.g., $\nabla R_{a,d^+}(rot(T \cup nK)) = \nabla R_{a,d^+}(rot(T \cup -flip(nK)))$ for each $n > 1$ and each $0 < a < n$.

The same is true of course also for $\bar{R}_a^2$ instead of $\nabla R_{a,d^+}$.

It is easy to see that all signs and all homological markings of crossings are invariant under the operation $-flip$. For a triple crossing p, the crossing d stays invariant under $-flip$. However, the crossings hm and ml interchange because the direction of the z-axis has changed. Remember that the symbol of the weights for $\nabla R_{a,d^+}$ is *always* defined with respect to hm, which becomes ml under $-flip$ and which hence does no longer contribute at all!

We show, e.g., changings of global types of triple crossings which appear in $\bar{R}_a^{(2)}$ in Fig. 2.20. It follows that the set of degenerate triple crossings in the formula, which contribute to $\bar{R}_a^{(2)}(rot(T \cup nK))$, is invariant under $-flip$. However, the weights change completely because each contributing couple $(n, 0)$ becomes a couple $(0, n)$ with ∞ between the two heads (and hence it does no longer contribute).

A combinatorial 1-cochain is for us an unordered set of integers associated with an oriented loop (the set is unordered because of the commutation relations (a), compare the Introduction), and the corresponding cohomology class is the sum of these integers. Because the weights which contribute to $\bar{R}_a^{(2)}(rot(T \cup nK))$ and those which contribute to $\bar{R}_a^{(2)}(rot(-flip(T \cup nK)))$ are completely different, we can expect that the corresponding 1-cocycles are completely different too. However, we do not know if the 1-cohomology class is different too because examples are too complicated to be calculated by hand.

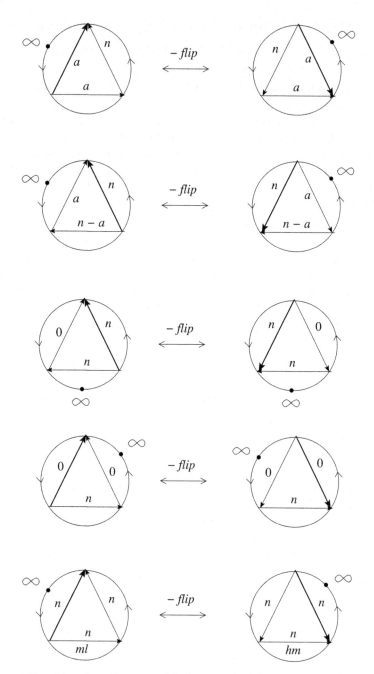

Figure 2.20: Changings for some global types of triple crossings under $-flip$.

In any case, we do already a little better than the usual quantum knot invariants as, e.g., the HOMFLYPT polynomial. We can see them as combinatorial 0-cohomology classes in the moduli space of knot diagrams. The corresponding combinatorial 0-cocycles would be the unordered set of polynomials which represent the ends of the branches of the tree, which corresponds to the calculation of the quantum invariant of a knot K by using skein relations, see, e.g., Ref. [31]. But this tree is identical to the tree for the calculation of the quantum invariant for the inverse knot $-K$. Hence, already the combinatorial 0-cocycles are the same and not only the 0-cohomology classes.

It is well known that quantum knot invariants are dominated by finite type invariants, compare, e.g., Ref. [2]. On the other hand, there aren't any finite type invariants known which could distinguish the orientations of classical knots. But there are such invariants already for 2-component string links, see Ref. [14]. The reason is simple: if we consider a crossing between the two components, then $-flip$ interchanges the overcross with the undercross and hence the two components. Consequently, the position of the self-crossings of the two components with respect to the crossing between the two components change in the invariant. In our case, the crossings hm and ml in triple crossings interchange already for knots. This makes it potentially possible that our 1-cocycles could perhaps distinguish the orientation for classical knots, if we are lucky.

Let $K = 8_{17}$ with its standard diagram in the knot atlas. First good candidates would be the refined 1-cocycles $\nabla R_{2_i,d^+}(fh(\sigma_1\sigma_2 \cup 3K))$ or $\bar{R}_{2_i}^{(2)}(fh(\sigma_1\sigma_2 \cup 3K))$ and, respectively, $\nabla R_{2_i,d^+}(fh(\sigma_1\sigma_2 \cup -flip(3K)))$ and $\bar{R}_{2_i}^{(2)}(fh(\sigma_1\sigma_2 \cup -flip(3K)))$, for $i = 1$ and $i = 2$. Another good candidate would be $R_{i,j,k}^{(2)}(fh(\sigma_1\sigma_2 \cup 3K))$ for the 3-cable with i,j,k all distinct because we wouldn't need then regular isotopy. If the values of the corresponding 1-cocycles are not equal, then we would have reproven that the knot 8_{17} is not invertible, but this time by using an universal knot invariant which can be calculated with polynomial complexity, and without making any explicit use of the knot group.

Notice that it is better to consider the 3-cable here, not only because of the splitting, but because 8_{17} is known to be isotopic to its mirror image with reversed orientation. The mirror image interchanges the markings a and $n - a$, which would be the same for $n = 2$ and $a = 1$. Moreover, $T = \sigma_1\sigma_2$ is not isotopic to $-flip(T) = \sigma_2\sigma_1$, which could perhaps be important too.

It would be of crucial importance for all the potential applications in this section, to have a computer program in order to calculate our 1-cocycles for lots of knots.

Chapter 3

Proofs

3.1 Generalities and Preparations

We use the technology which was developed in Ref. [16]. For the convenience of the reader, we recall here the main lines of our approach.

We study the discriminant Σ of non-generic diagrams in $\tilde{M}_n$ or M_n together with its natural stratification. Our strategy is the following: for an oriented generic loop in $\tilde{M}_n$ or M_n, we associate an integer to the intersection with each stratum in $\Sigma_{\text{tri}}^{(1)}$, i.e., to each Reidemeister III move, and we sum up over all moves in the loop.

In order to show that our 1-cochains are 1-cocycles, we have to prove that the sum is 0 for each meridian of strata of codimension 2, i.e., in $\Sigma^{(2)}$. This is very complex, but we have used strata from $\Sigma^{(3)}$ in order to reduce the proof to a few strata in $\Sigma^{(2)}$. It follows that our sum is invariant under generic homotopies of loops in $\tilde{M}_n$ or M_n. But it takes its values in an abelian ring, and hence, it is a 1-cocycle. Showing that the 1-cocycle is 0 on the meridians of the quadruple crossings $\Sigma_{\text{quad}}^{(2)}$ is by far the hardest part. This corresponds to finding a new solution of the *tetrahedron equation*.

Consider four oriented straight lines which form a braid such that the intersection of their projection into $\mathbb{C}$ consists of a single point. We call this an *ordinary quadruple crossing*. After a generic perturbation of the four lines, we will now exactly see six ordinary crossings. We assume that all six crossings are positive, and we call the corresponding quadruple crossing a *positive quadruple crossing*. Quadruple crossings form smooth strata of codimension 2 in the topological moduli space of lines in 3-space which is equipped with a fixed projection pr. Each generic point in such a stratum is adjacent to exactly eight smooth strata of codimension 1. Each of them corresponds to configurations of lines which have exactly one ordinary triple crossing besides the remaining ordinary crossings. We number the lines from 1 to 4 from the lowest to the highest (with respect to the projection pr).

The eight strata of triple crossings glue pairwise together to form four smooth strata which intersect pairwise transversely in the stratum of the quadruple crossing, see Fig. 3.1 and compare Refs. [16,19]. The strata of triple crossings are determined by the names of the three lines which give the triple crossing. For shorter writing, we give them names from P_1 to P_4 and $\bar{P}_1$ to $\bar{P}_4$ for the corresponding

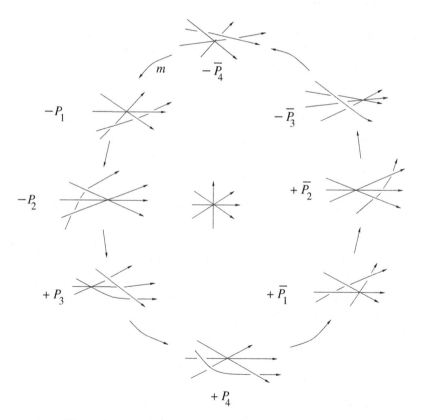

Figure 3.1:   Unfolding of a positive quadruple crossing.

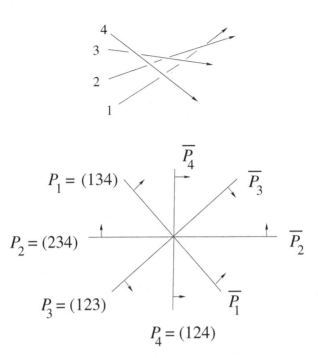

Figure 3.2:   The names of the strata of triple crossings in the intersection of a normal 2-disc of a positive quadruple crossing.

stratum on the other side of the quadruple crossing. We show the intersection of a normal 2-disc of the stratum of codimension 2 of a positive quadruple crossing with the strata of codimension 1 in Fig. 3.2. The strata of codimension 1 have a natural coorientation, compare Definition 2.2. We could interpret the six ordinary crossings as the edges of a tetrahedron and the four triple crossings likewise as the vertices or the 2-faces of the tetrahedron. For the classical tetrahedron equation, one associates to each stratum P_i, i.e., to each vertex or equivalently to each 2-face of the tetrahedron, some operator (or some R-matrix) which depends only on the names of the three lines and to each stratum $\bar{P}_i$ the inverse operator. The tetrahedron equation now says that if we go along the meridian, then the composition of these operators is equal to the identity. Note that in the literature, see, e.g., Ref. [30], one considers planar configurations of lines. But this is of course equivalent to our situation because all crossings are positive and hence the lift of the lines into 3-space is

determined by the planar picture. Moreover, each move of the lines
in the plane which preserves the transversality lifts to an isotopy of
the lines in 3-space. The tetrahedron equation has many solutions;
the first one was found by Zamolodchikov, see, e.g., Ref. [30].

However, the solutions of the classical tetrahedron equation are
not well adapted in order to construct 1-cocycles for moduli spaces of
knots. A local solution of the tetrahedron equation is of no use for us
because as already pointed out there are no integer-valued 1-cocycles
for all knots in the 3-sphere. We have to replace them with long knots
or more generally with points in M_n. For knots in the solid torus V^3,
we can now associate with each crossing in the diagram a winding
number (i.e., a homology class in $H_1(V^3)$) in a canonical way. Therefore, we have to consider six different positive tetrahedron equations,
corresponding to the six different abstract closures of the four lines
to a circle, and in each of the six cases, we have to consider all possible winding numbers of the six crossings. We call this the *positive
global tetrahedron equations*. There are exactly six global types of
positive quadruple crossings without the homological markings. We
show them in Fig. 3.3.

One easily sees that there are exactly 48 local types of quadruple
crossings (analogue to the eight local types of triple crossings).

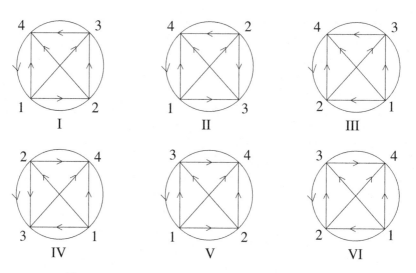

Figure 3.3:   The global types of quadruple crossings.

We study the relations of the local types of triple crossings in what we call the *cube equations*. Triple crossings come together in points of $\Sigma^{(2)}_{\text{trans}-\text{self}}$, i.e., an auto-tangency with an addition transverse branch [16]. But one easily sees that the global type of the triple crossings (i.e., its Gauss diagram with the homological markings but without the writhe) is always preserved. We now make a graph Γ for each global type of a triple crossing in the following way: the vertices correspond to the different local types of triple crossings. We connect two vertices by an edge if and only if the corresponding strata of triple crossings are adjacent to a stratum of $\Sigma^{(2)}_{\text{trans}-\text{self}}$. We have shown that the resulting graph is the 1-skeleton of the 3-dimensional cube I^3, see Fig. 3.4. In particular, it is connected. The edges of the graph $\Gamma = skl_1(I^3)$ correspond to the types of strata in $\Sigma^{(2)}_{\text{trans}-\text{self}}$. The solution of the positive tetrahedron equation tells us what is the contribution to the 1-cocycle of a positive triple crossing (i.e., all three involved crossings are positive). The meridians of the strata from $\Sigma^{(2)}_{\text{trans}-\text{self}}$ give equations which allow us to determine the contributions of all other types of triple crossings. However, a global phenomenon occurs: each loop in Γ could give an additional equation. Evidently, it suffices to consider the loops which are the boundaries of the 2-faces from $skl_2(I^3)$. We call all the equations which come from the meridians of $\Sigma^{(2)}_{\text{trans}-\text{self}}$ and from the loops in $\Gamma = skl_1(I^3)$ the *cube equations*.

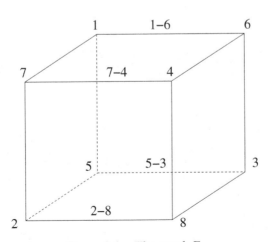

Figure 3.4: The graph Γ.

(Note that a loop in Γ is more general than a loop in $\tilde{M}_n$ or M_n. For a loop in Γ, we come back to the same local type of a triple crossing but not necessarily to the same whole diagram of the knot.)

Our strategy is the following: first we find a solution of the positive global tetrahedron equation. Then, we solve the cube equations by adding *correction terms* for different local types of triple crossings but which vanish for positive triple crossings. We have shown in Ref. [16] that the resulting 1-cochain is then a 1-cocycle in M_n.

For the convenience of the reader, we give here again the figures which are the main tools for our work, see Ref. [16]. Let us consider the global positive quadruple crossings. We naturally identify crossings in an isotopy outside Reidemeister moves of types I and II. The Gauss diagrams of the unfolding of the quadruple crossings are given in Figs. 3.5–3.16. We marque also the different possibilities for the point at infinity in the case of long knots in the solid torus. The (positive) crossing between the local branch i and the local branch j is always denoted by ij. We give then the homological markings, which depend on three parameters α, β and γ on the circle.

Using these figures, we show in Figs. 3.17–3.28 the homological markings of the arrows. Here, α, β and γ are the homology classes represented by the corresponding arcs in the circle (remember that the circle in the plan is always oriented counterclockwise). The ordering of the global types here is more adapted to the check for our 1-cocycles.

In M_n, we have the unknown parameters α, β and γ in $\{0, 1, \ldots, n\}$ and $\alpha + \beta + \gamma \leq n$ because there are no negative loops in the diagrams.

3.2   Proof of Theorems 2.1–2.4

All five 1-cochains which were defined in Chapter 2 satisfy the commutation relations (a).

Indeed, we have only to check that the contribution of a triple crossing p does not change if we pass simultaneously through another Reidemeister move p'. If p' is an R II move, then evidently the contribution does not change because the two new crossings do never contribute together in a configuration and their individual contributions cancel out because the two crossings have different signs. If p' is

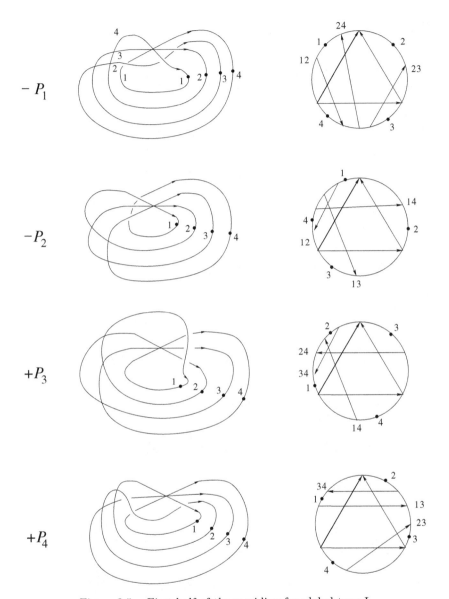

Figure 3.5: First half of the meridian for global type I.

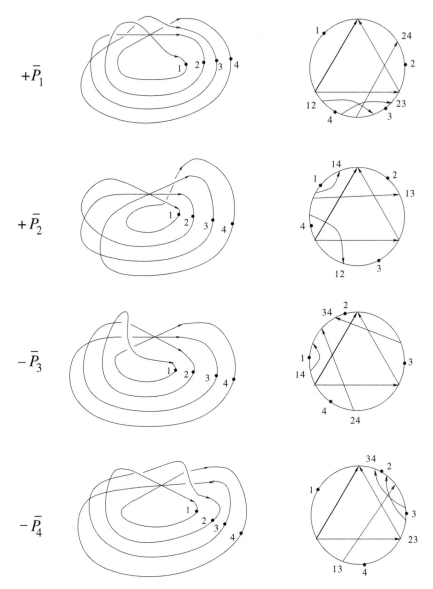

Figure 3.6:   Second half of the meridian for global type I.

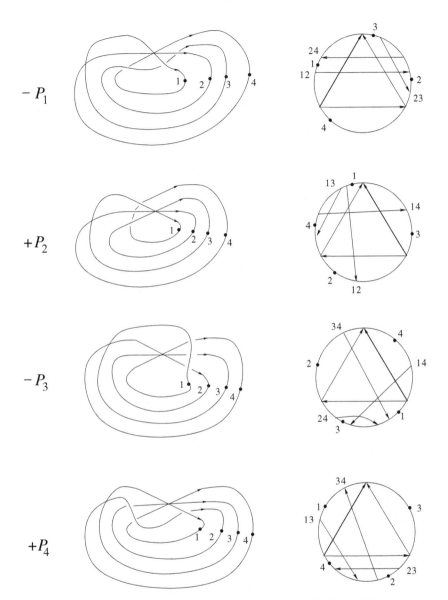

Figure 3.7: First half of the meridian for global type II.

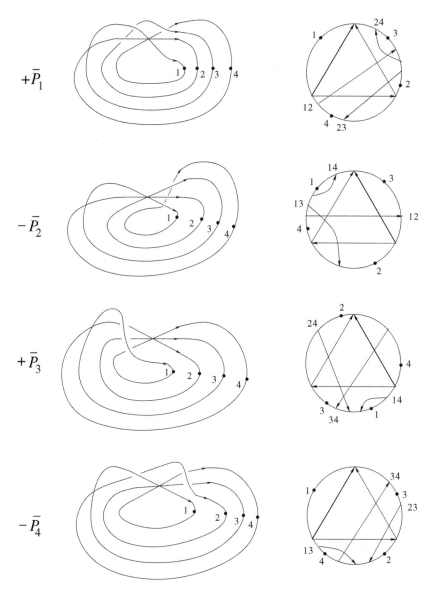

Figure 3.8:   Second half of the meridian for global type II.

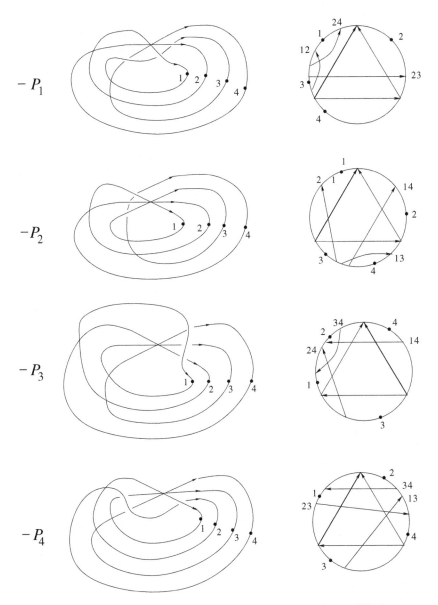

Figure 3.9: First half of the meridian for global type III.

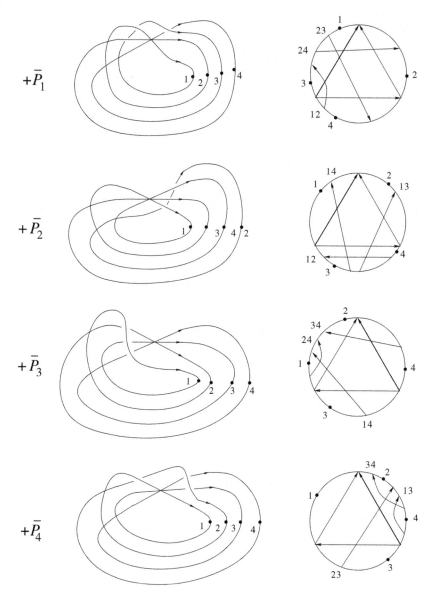

Figure 3.10:   Second half of the meridian for global type III.

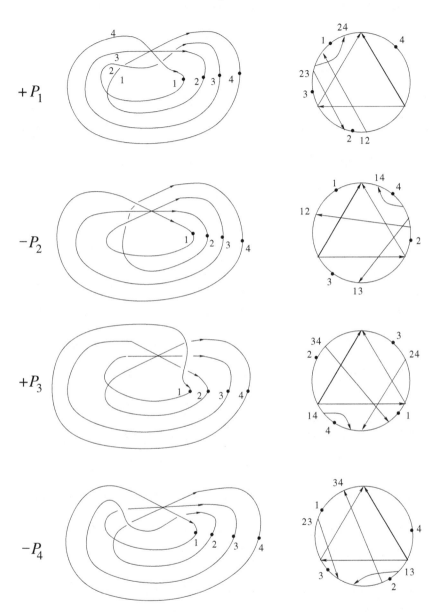

Figure 3.11:   First half of the meridian for global type IV.

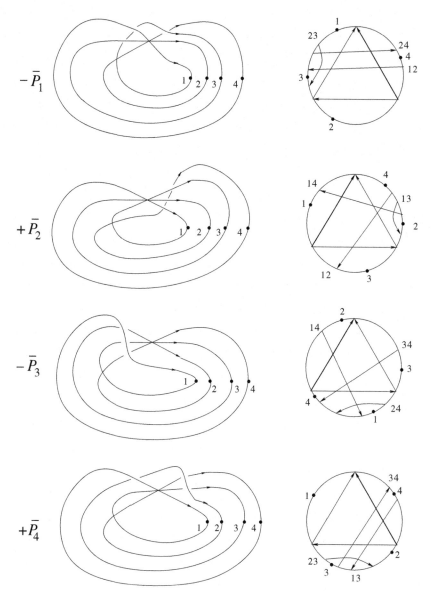

Figure 3.12:   Second half of the meridian for global type IV.

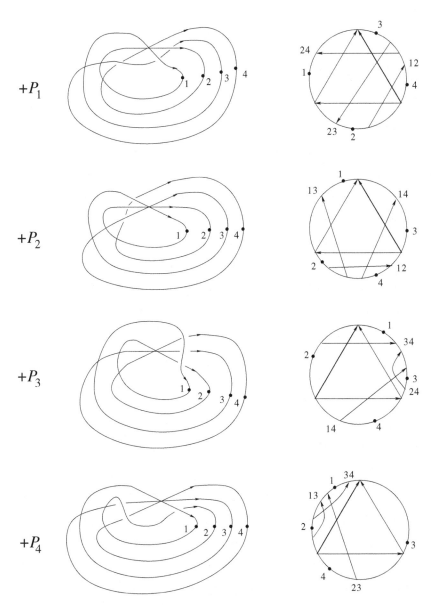

Figure 3.13:   First half of the meridian for global type V.

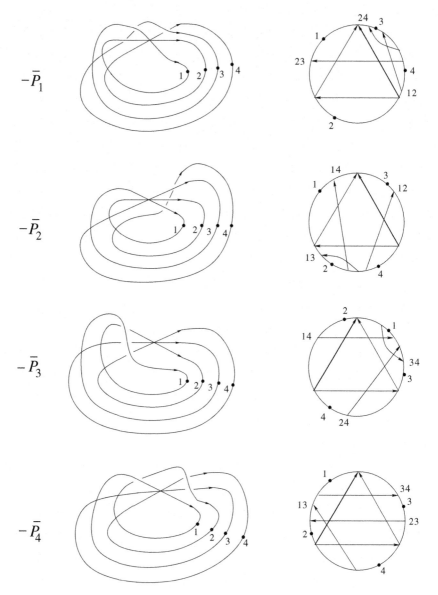

Figure 3.14:   Second half of the meridian for global type V.

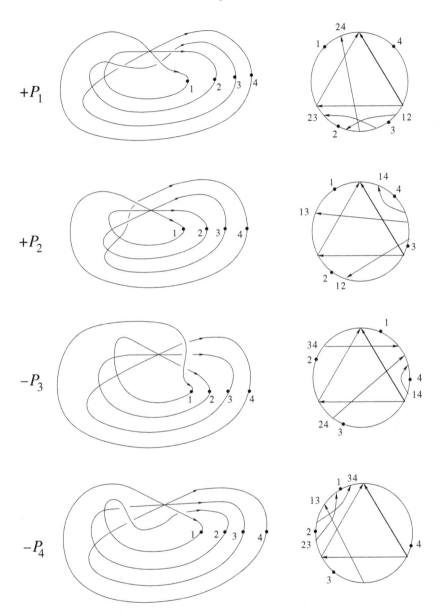

Figure 3.15: First half of the meridian for global type VI.

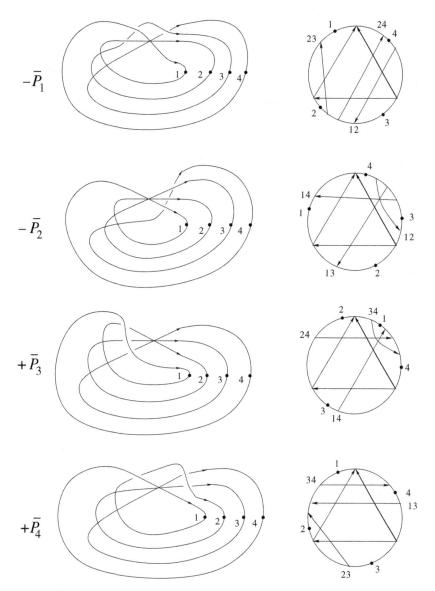

Figure 3.16:   Second half of the meridian for global type VI.

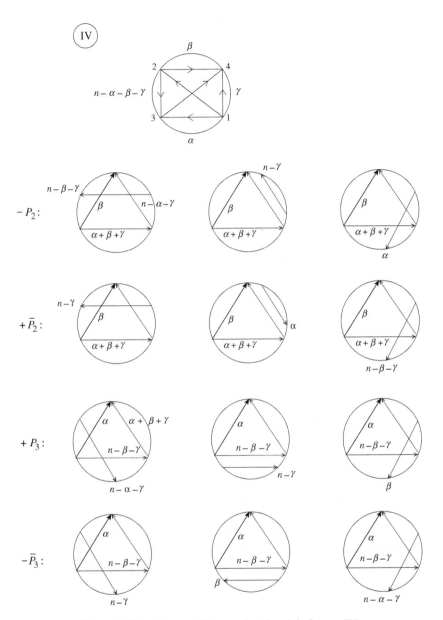

Figure 3.17:   Moves of type r for the global type IV.

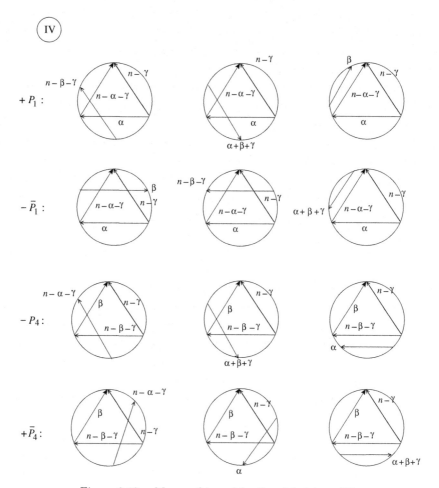

Figure 3.18:   Moves of type l for the global type IV.

an R I move, then the new crossing is isolated in the Gauss diagram and hence does not contribute neither to a configurations C_{2k} nor to the refined linking numbers $l(p)$.

As usual, because the commutation relations are satisfied for R II moves, it is sufficient to prove it only for the local type of positive triple crossings p' (i.e., all three crossings are positive) because of the cube equations. If p' is an R III move, then the refined linking numbers $l(p)$, $l_a(p)$ and $l_{n-a}(p)$ stay invariant as weights of order one, i.e., defined by single crossings. Moreover the relative weights

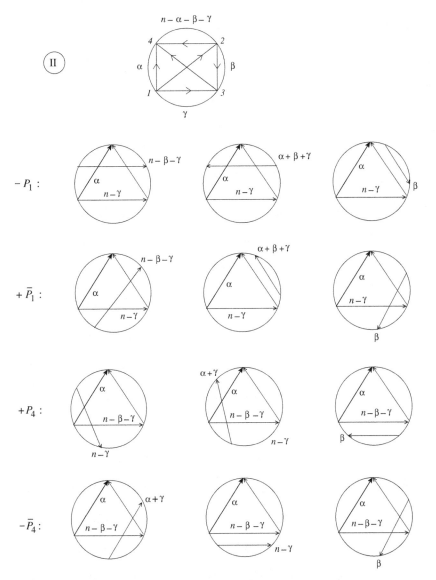

Figure 3.19: Moves of type r for the global type II.

$W_{2k-1}(hm)$ and $\bar{W}_{2k-1}(hm)$ stay invariant. Indeed, we have to consider only degenerate triple crossings p' because the weight could change only if at least two of the three crossings in p' contribute to the weight. The contributions to C_{2k} are invariant under passing p'

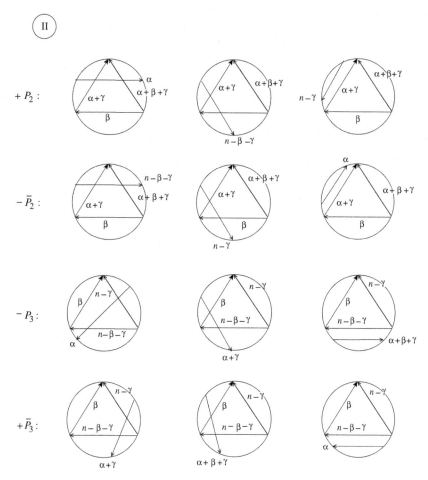

Figure 3.20: Moves of type l for the global type II.

because C_{2k} is a finite type invariant (the coefficient of z^{2k} in the Conway polynomial). Hence, the relative weight could only change if an n-crossing from p' now becomes the first n-crossing in the configuration instead of the crossing hm in p. This implies that the foot of the n-crossing is nearer to ∞ than the foot of hm. But this is not possible. The foot of the n-crossing in p' coincides with the foot or the head of a crossing q in p'. If hm was the first n-crossing in a configuration C_{2k} before the move p', then neither the n-crossing nor the crossing q could contribute to W_{2k-1}. Consequently, only

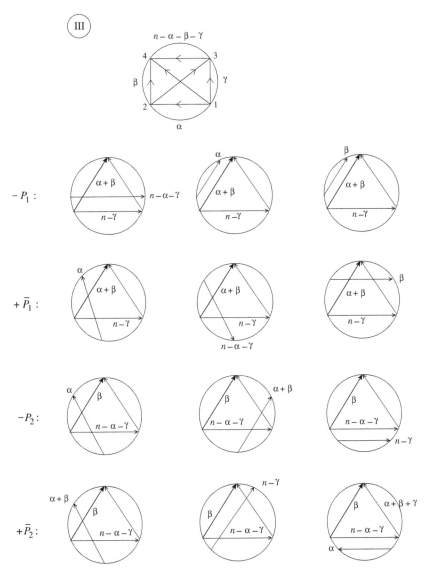

Figure 3.21: Moves of type r for the global type III.

the remaining crossing from p' could contribute before and after the move and W_{2k-1} is evidently invariant.

In the case of $\bar{W}_{2k-1}(hm)$, we have to also consider the possibility that the last 0-crossing before p' is no longer the last 0-crossing

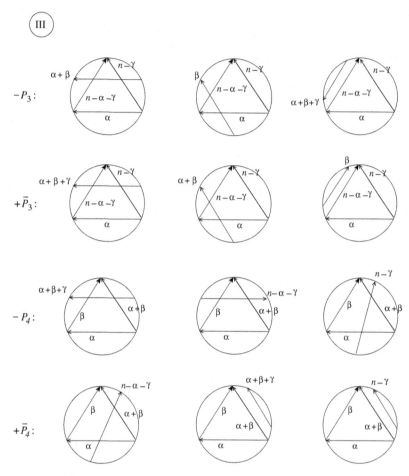

Figure 3.22: Moves of type l for the global type III.

after p'. But the considerations are exactly the same as for the weights $\bar{W}_{2k}(p)$, which we will consider now.

Let an n-crossing c from the three crossings of p' be the first n-crossing in $\bar{W}_{2k}(p)$ before p' and let us assume that c is no longer the first n-crossing in $\bar{W}_{2k}(p)$ after p'. The following are the two possibilities:

Case 1

The crossing c has the same foot as a crossing q in p' which is also an n-crossing. In this case, p' is necessarily of type $r(n,n,n)$ or $l(n,0,n)$.

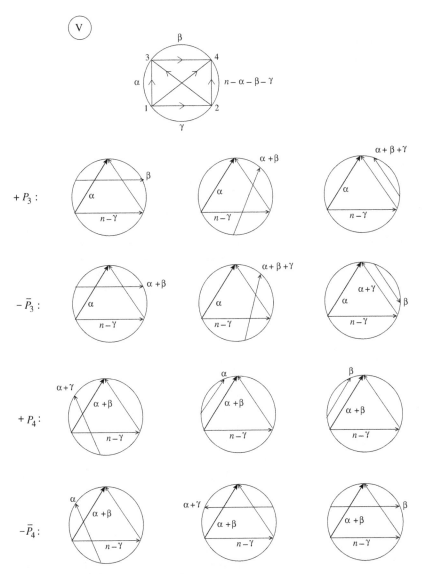

Figure 3.23: Moves of type r for the global type V.

No matter with which crossings of p' the crossing c contributes to C_{2k} before p', after p' either c or q or both contribute to C_{2k}, and hence, in any case, the first foot of an n-crossing is still in hm^+ and consequently the contribution of C_{2k} is still in $\overline{W}_{2k}(p)$.

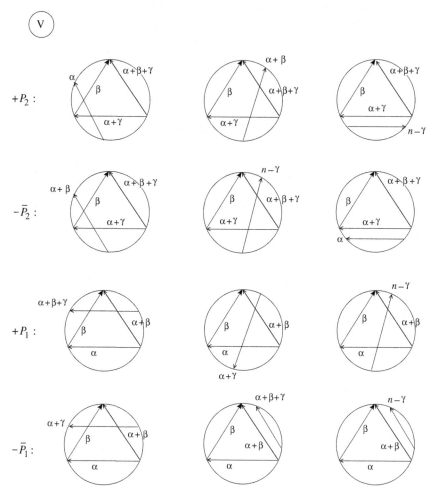

Figure 3.24:   Moves of type l for the global type V.

Case 2

The foot of c is not the foot of another n-crossing in p'. If there
is exactly only one n-crossing in p', then c contributes before and
after p'. Indeed, if the two 0-crossings after p' contribute together
and c does not contribute, then there should be another n-crossing
outside p' which is the first n-crossing in the configuration because of
the first–last foot property of each configuration in C_{2k} (compare the
Introduction). But then c was not the first n-crossing. If c contributes

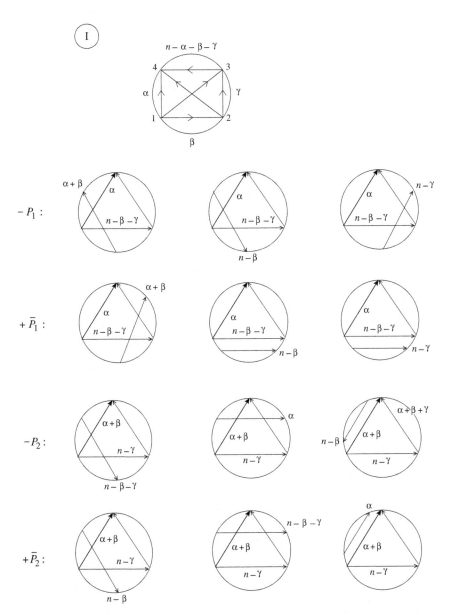

Figure 3.25: Moves of type r for the global type I.

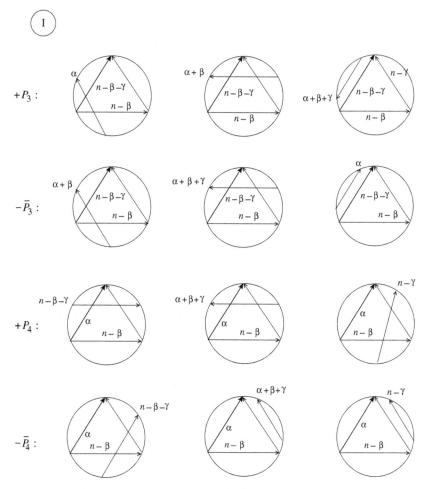

Figure 3.26:   The remaining moves of type r for the global type I.

after p' with one of the 0-crossings, then either c is still the first
n-crossing or the head or the foot of the 0-crossing is before the
foot of c. Then again, by the first–last foot property, there has to be
another n-crossing before c, and hence, c was not the first n-crossing.

Consequently, it remains to study the cases where at least two
crossings of p' are n-crossings. If all three are n-crossings, then p'
is of type $r(n, n, n)$. If at least two of the n-crossings contribute,
then we are either in Case 1 or on one side of p' the crossing ml contributes with hm and on the other side of p' the crossing d contributes

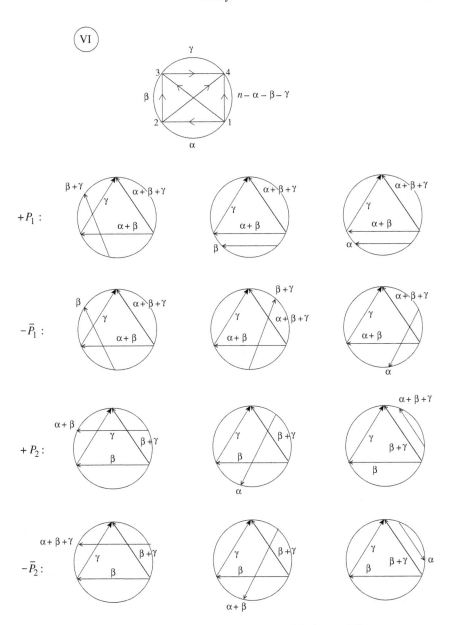

Figure 3.27: Moves of type *l* for the global type VI.

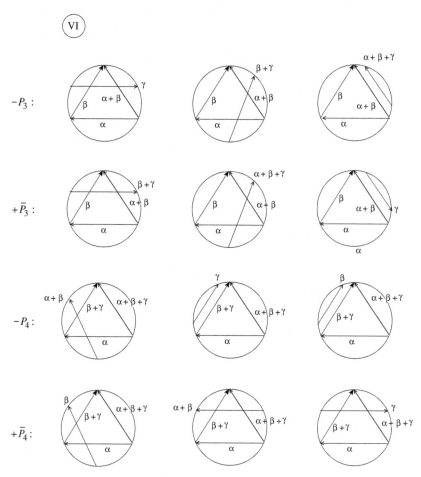

Figure 3.28: The remaining moves of type l for the global type VI.

with hm. Consequently, ml and d are the first n-crossings, respectively. But they have the same foot in hm^+.

If exactly two crossings in p' are n-crossings, then p' is of type $l(n,0,n)$ or of type $l(n,n,0)$. The type $l(n,0,n)$ was already considered in Case 1. It remains to consider the type $l(n,n,0)$. We give the two sides of p' in Fig. 3.29.

On the negative side, the crossing hm could be the first n-crossing and it could contribute with the crossing ml or with the crossing d or with both. If it contributes with ml alone, then on the positive

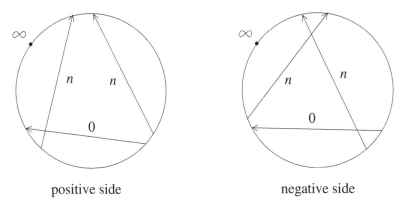

positive side   negative side

Figure 3.29:   The two sides of $l(n, n, 0)$.

side it contributes with d alone because of the preserving intersection property (compare the Introduction). In this case, it stays the first n-crossing. If it contributes with d alone, then on the positive side it contributes with ml again because of the preserving intersection property. But then hm could not be the first n-crossing on neither side of p' because of the first–last foot property. If all three contribute on one side, then they contribute all three also on the other side and again hm could not be the first n-crossing on neither side of p' because of the first–last foot property.

Completely analogous considerations work also for the last 0-crossings. We left the verification to the reader.

We have shown that all our combinatorial 1-cochains, including those from Conjecture 2.1, satisfy the commutation relations (a).

The 1-cochains $R^{(2k)}_{a,d^+,hm}$ and $R^{(2k)}_{a,d^+}$ satisfy the positive global tetrahedron equation (b).

This is the heart of the proof.

We have given the homological markings, which depend on three parameters α, β and γ on the circle in Figs. 3.17–3.28. Here, α, β and γ are the homology classes represented by the corresponding arcs in the circle (remember that the circle in the plan is always oriented counterclockwise).

We have to show now that $R^{(2k)}_{a,d^+,hm}$ and $R^{(2k)}_{a,d^+}$ without the term $(w(hm) - 1)W_{2k-1}(hm)w(hm)$ in $r(a, n, a)$ (which vanishes automatically for positive triple crossings) vanishes on the meridian of each global positive quadruple crossing, i.e., it satisfies the positive global tetrahedron equation.

The eight strata in the meridian come in pairs with different signs: P_i and $\bar{P}_i$. The crossings d, hm and ml are of course the same for P_i and $\bar{P}_i$. These two strata differ with respect to the triangle by exactly the three remaining crossings from the quadruple crossing. But only n-crossings with the foot in hm^+ could contribute to $W_{2k}(p)$ and only n-crossings with the foot in hm^- could contribute to $W_{2k-1}(hm)$. We have therefore to study when the foot of an n-crossing has slide over the foot or the head of the crossing hm from P_i with respect to $\bar{P}_i$.

Inspecting Figs. 3.5–3.16, we see that the crossings, for which the foot slides over hm, do not depend only on the global type (i.e., I up to VI) of the quadruple crossing. Remember that we call c an f-crossing if its homological marking $[c] = n$ and the foot of c is in the open arc hm^+. We have to study only the following crossings:

P_3: foot of 24 in $hm^-(hm = 23)$, $\bar{P}_3$: foot of 24 in hm^+

P_3: foot of 34 in $hm^+(hm = 23)$, $\bar{P}_3$: foot of 34 in hm^-

P_4: foot of 23 in $hm^+(hm = 24)$, $\bar{P}_4$: foot of 23 in hm^-.

Note, that this never happens for P_1 and $\bar{P}_1$ and for P_2 and $\bar{P}_2$. Indeed, for P_2 and $\bar{P}_2$, the foots stay all in the same arc, and for P_1 and $\bar{P}_1$, a foot can only slide over a foot of ml and hence it stays in hm^+.

Remember that the edges of the tetrahedron correspond to the crossings and the 2-faces correspond to the triple crossings. If there is one 2-face corresponding to $r(a, n, a)$ or $r(n, n, n)$, then the remaining three edges have a common vertex. The homology class of one of these edges (i.e., the homological marking of the corresponding crossing) determines now the homology classes of the remaining two edges. Of course, the homology classes of three edges with a common vertex determine the homology classes of the remaining three edges.

As already mentioned, we call a tetrahedron *generic* (with respect to a) if there is at least one stratum of global type $r(a, n, a)$ or if it contains one stratum of triple crossings with all markings in $\{0, n\}$ and there is at least one crossing of marking a (there are then automatically two other crossings with markings in $\{a, n - a\}$ and there are no other triple crossings with all markings in $\{0, n\}$).

We call the tetrahedron *degenerate* if all six crossings have their markings in $\{0, n\}$.

Our strategy is the following: we solve the generic tetrahedron equations by using weights of order $2k$. We take then this solution and possibly complete it to a solution of the degenerate tetrahedron equations.

Let us start with the simpler case of $R_{a,d^+,hm}^{(2k)}$.

In each of the six global types, we have now to consider how $W_{2k-1}(hm)$ changes and how this changing is compensated by a configuration with $W_{2k-1}(hm)$ for another stratum. We do this by inspecting simultaneously a figure from Figs. 3.5 to 3.16 and the corresponding figure from Figs. 3.17 to 3.28.

Using Figs. 3.17–3.28, we go through the list of all cases where at least one of the strata P_i is of the global type $r(a, n, a)$ (or of type $r(n, n, n)$, which will be needed only for $R_{a,d^+}^{(2k)}$), and moreover, there is at least one other n-crossing or one other 0-crossing, and hence, the weights could change.

This gives us conditions on the parameters α, β and γ. The weight $W_{2k-1}(hm)$ can only change if there is at least one other stratum with the three markings in $\{0, n\}$. If P_i does not have the right global type in order to contribute or if $W_{2k-1}(hm)(P_i) = W_{2k-1}(hm)(\bar{P}_i)$, then we will simply skip them in the proof (because always $sign(P_i) = -sign(\bar{P}_i)$ and they cancel out together).

Let us begin with the following important definition.

Definition 3.1 *Let c and c' be two persistent crossings of the tetrahedron (i.e., two edges of the tetrahedron with homological markings 0 or n). Then, $X(c, c')$ is the sum of those contributions to $W_{2k}(p)$, respectively, $W_{2k-1}(hm)$, if $c = hm$, where c and c' intersect (i.e., the corresponding chords in the Gauss diagram intersect) and c or c' is the first n-crossing, and correspondingly $II(c, c')$ is the sum of those contributions where c and c' do not intersect and c is the first n-crossing.*

Global type I

All global types of triple crossings in the meridian are of type r in this case.

I_1: $\alpha = a$, $\beta = n - a$ and hence $\gamma = 0$. Then, P_1, P_3 and P_4 are of type $r(a, n, a)$ and P_2 is of type $r(n, n, n)$. In this case, 23, 24 and 34 are all n-crossings.

$-P_1 + \bar{P}_1$: $\infty = 1$, 23 and 24 cannot contribute to $W_{2k-1}(34)$ because 34 has to be the first n-crossing. Consequently, $-W_{2k-1}(34)(P_1)$ and $W_{2k-1}(34)(\bar{P}_1)$ cancel out together.

$P_3 - \bar{P}_3$: If $\infty = 2$, then $W_{2k-1}(23)(P_3)$ and $-W_{2k-1}(23)(\bar{P}_3)$ cancel out together. Indeed, 24 and 34 cannot contribute because their heads would be in the arc from ∞ to the foot of $hm = 23$.

$\infty = 1$: In this case, 23, 24 and 34 could contribute together for P_3 and they would no longer contribute together in $-\bar{P}_3$ because of the foot of 24. However, they could contribute together also in $-\bar{P}_4$, and they could not in P_4 because of the foot of 23. But their configurations in P_3 and in $-\bar{P}_4$ correspond exactly to the triple crossing $P_2 = \bar{P}_2 = 234$ before and after the R III move. Consequently, by the preserving intersection property (compare the Introduction), they give the same contribution to $R^{(2k)}_{a,d^+,hm}$ (no matter what the remaining crossings in the configurations in C_{2k} are) and they cancel out together.

It remains to consider the configurations which contain hm and exactly one other n-crossing from the tetrahedron.

$P_3 - \bar{P}_3$: $W_{2k-1}(hm)(P_3) - W_{2k-1}(hm)(\bar{P}_3)$ is equal to $+II(23, 34) + X(23, 24) - X(23, 34)$. On the other hand, $P_4 - \bar{P}_4$ contributes $X(24, 34) - II(24, 34) - II(23, 24)$ to $W_{2k-1}(hm)(P_4) - W_{2k-1}(hm)(\bar{P}_4)$.

We apply now the preserving intersection property to the R III move which corresponds to $P_2 = \bar{P}_2 = 234$. For the convenience of the reader, we give the two sides of the move in Fig. 3.30. We obtain $X(23, 34) = X(23, 24) + X(24, 34)$ and $II(23, 34) = II(23, 24) + II(24, 34)$. Consequently, $R^{(2k)}_{a,d^+,hm} = 0$ on the meridian of the quadruple crossing in this case.

I_2: $\alpha = a$, $\beta = 0$ and $\gamma = n - a$. Then, P_1 and P_2 are of type $r(a, n, a)$ and P_3 and P_4 are of type $r(a, a, n)$. Only 12 is an

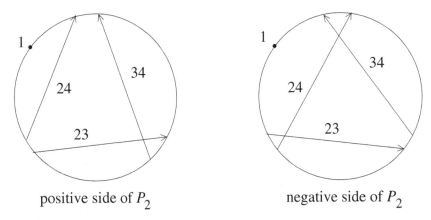

positive side of P_2 negative side of P_2

Figure 3.30:   The two sides of P_2 of type $r(n, n, n)$.

n-crossing, but its position with respect to $hm = 34$ does never change because 12 and 34 cannot be together in a triple crossing for the tetrahedron. It follows that $R^{(2k)}_{a,d^+,hm} = 0$ on the meridian of the quadruple crossing in this case.

I_3: $\alpha = 0$, $\beta = a$ and $\gamma = n - a$. Then, P_2 is of type $r(a, n, a)$ and P_1 is of type $r(0, n, 0)$. There are no other n-crossings, and hence, $hm = 34$ stays the first n-crossing in both $-P_2$ and $\bar{P}_2$. The two 0-crossings 13 and 14 together with 34 form the triple crossing $P_1 = \bar{P}_1 = 134$. Consequently, $-W_{2k-1}(hm = 34)(P_2) + W_{2k-1}(hm = 34)(\bar{P}_2) = 0$ because C_{2k} is invariant under R III moves.

I_4: $\alpha = n - a$, $\beta = a$ and hence $\gamma = 0$. Then, P_2 is of type $r(n, n, n)$, but there is no P_i of type $r(a, n, a)$ at all.

I_5: $\alpha = n$ and hence $\beta = 0$ and $\gamma = 0$. Then, all four P_i are of type $r(n, n, n)$.

I_6: $\alpha = 0$, $\beta = n$ and hence $\gamma = 0$. Then, P_2 is of type $r(n, n, n)$ and there are no $r(a, n, a)$.

I_7: $\alpha = $ arbitrary, $\beta = 0$ and $\gamma = 0$. Then, P_3 is of type $r(n, n, n)$ but again there are no $r(a, n, a)$.

Global type II
Here, only the strata P_1 and P_4 could contribute.

II_1: $\alpha = a$, $\beta = 0$ and $\gamma = n - a$. Then, P_1 and P_4 are of type $r(a, n, a)$ and P_2 is of type $l(n, n, 0)$ (the remaining P_3 is of type $l(a, 0, a)$ and need not to be considered at all). We only need to consider $\infty = 1$. A priory 23, 24 and $hm = 34$ could contribute together in a configuration in A_{2k} which would contribute to $-W_{2k-1}(hm = 34)(P_1)$ because hm is the first n-crossing, and it would not contribute to $W_{2k-1}(hm = 34)(\bar{P}_1)$ because the head of 23 is now before the foot of hm. However, the three crossings form the triple crossing P_2. If they contribute together to a configuration A_{2k} on one side of P_2, then they contribute also to a (different) configuration A_{2k} on the other side because of the preserving intersection property. But on the positive side of P_2, we come from $\infty = 1$ first to the head of the 0-crossing 23. It follows now from the first–last foot property (compare the Introduction) that there should be another n-crossing in A_{2k} (invisible, because it is not a crossing from the tetrahedron) which has its foot before the head of 23. But in this case, the configurations A_{2k} which contain 23, 24 and $hm = 34$ do not contribute to $-W_{2k-1}(hm = 34)(P_1)$ neither because hm is not the first n-crossing.

One easily sees that 23, 34 and $hm = 24$ can never contribute together in neither P_4 nor $\bar{P}_4$ because hm is evidently never the first n-crossing.

We consider now the configurations of hm with exactly one other persistent crossing of the tetrahedron.

$-P_1$: contribution $-X(34, 24) + II(34, 23)$
$\bar{P}_1$: contribution $II(34, 24)$.

We give the two sides of the R III move P_2 in Fig. 3.31.
The configurations $X(34, 24)$ become the configurations $X(34, 23)$ on the other side. The same considerations as in II_1 show now that there is an invisible first n-crossing and hence there are no contributions of $X(34, 24)$ to $-W_{2k-1}(hm = 34)(P_1)$. Configurations from $II(34, 23)$ correspond on the other side to $II(34, 24)$ or

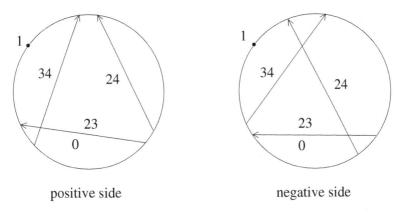

positive side   negative side

Figure 3.31:   The two sides of P_2 of type $l(n, 0, n)$ for the global type II.

to $II(23, 24)$. In the first case, we get $-W_{2k-1}(hm = 34)(P_1) + W_{2k-1}(hm = 34)(\bar{P}_1) = 0$. In the second case, there has to be again an invisible first n-crossing for $II(23, 24)$ and consequently $II(34, 23)$ does not contribute to $-W_{2k-1}(hm = 34)(P_1)$ at all.

II_2: $\alpha = a$, $\beta = n - a$ and hence $\gamma = 0$. Then, P_4 is of type $r(a, n, a)$, P_1 is of type $r(a, a, n)$, only 13 and 24 are persistent crossings, and they are not together in a triple crossing. It follows that $W_{2k-1}(hm = 24)(P_4) - W_{2k-1}(hm = 24)(\bar{P}_4) = 0$.

II_3: $\alpha = n$, $\beta = 0$ and $\gamma = 0$. Then, P_1 and P_4 are of type $r(n, n, n)$ and there is no triple crossing of type $r(a, n, a)$ at all.

Global type III
Here, only the strata P_1 and P_2 could contribute.

III_1: $\alpha = a$, $\beta = 0$ and $\gamma = n-a$. Then, P_1 is of type $r(a, n, a)$ and P_2 is of type $r(0, n, 0)$. Since 23 and 24 are 0-crossings, the crossing $hm = 34$ always stays the first n-crossing. The three crossings 23, 24 and $hm = 34$ could contribute only simultaneously for $-P_1$ and $\bar{P}_1$. Again, because of $X(23, 34) = X(23, 24) + X(24, 34)$ and $II(23, 34) = II(23, 24) + II(24, 34)$ for $P_2 = \bar{P}_2 = 234$, we obtain $-W_{2k-1}(hm = 34)(P_1) + W_{2k-1}(hm = 34)(\bar{P}_1) = 0$.

III_2: $\alpha = 0$, $\beta = a$ and $\gamma = n - a$. Then, P_1 and P_2 are of type $r(a, n, a)$ and only 12 is a persistent crossing, namely, a 0-crossing.

It follows that $-W_{2k-1}(hm = 34)(P_1) + W_{2k-1}(hm = 34)(\bar{P}_1) = 0$ and that $-W_{2k-1}(hm = 34)(P_2) + W_{2k-1}(hm = 34)(\bar{P}_2) = 0$.

III_3: $\alpha = n-a$, $\beta = a$ and $\gamma = 0$. Then, P_1 is of type $r(n,n,n)$ and P_2 is of type $r(a,n,a)$. Since 13 and 14 are n-crossings with their foots always before the foot of 34, none of them could contribute neither to $-W_{2k-1}(hm = 34)(P_2)$ nor to $W_{2k-1}(hm = 34)(\bar{P}_2)$.

III_4: $\alpha = a$, $\beta = n - a$ and $\gamma = 0$. Then, P_1 is of type $r(n,n,n)$ and there is no stratum of type $r(a,n,a)$ at all.

Global type IV
Here, only the strata P_2 and P_3 could contribute.

IV_1: $\alpha = 0$, $\beta = a$ and $\gamma = 0$. Then, P_2 is of type $r(a,n,a)$. The only other persistent crossings are 13, which is an 0-crossing, and 14, which is an n-crossing. The foots of the crossings never change the arc for $-P_2$ and $\bar{P}_2$. Again, because of $X(13,34) = X(13,14) + X(14,34)$ and $II(13,34) = II(13,14) + II(14,34)$ for $P_1 = \bar{P}_1 = 134$, we obtain $-W_{2k-1}(hm = 34)(P_2) + W_{2k-1}(hm = 34)(\bar{P}_2) = 0$.

IV_2: $\alpha = a$, $\beta = n - a$ and $\gamma = 0$. Then, P_3 is of type $r(a,n,a)$. Only 14 is an n-crossing and it follows that $W_{2k-1}(hm = 23)(P_3) - W_{2k-1}(hm = 23)(\bar{P}_3) = 0$.

IV_3: $\alpha = a$, $\beta = 0$ and $\gamma = n - a$. Then, P_3 is of type $r(a,n,a)$. 24 and 34 are 0-crossings and they form with $hm = 23$ the triple crossing P_2. It follows from $X(23,34) = X(23,24) + X(24,34)$ and $II(23,34) = II(23,24) + II(24,34)$ for $P_2 = \bar{P}_2 = 234$ that $W_{2k-1}(hm = 23)(P_3) - W_{2k-1}(hm = 23)(\bar{P}_3) = 0$.

IV_4: $\alpha = 0$, $\beta = n$ and $\gamma = 0$. Then, P_2 is of type $r(n,n,n)$ and there is no stratum of type $r(a,n,a)$ at all.

Global type V
Here, only the strata P_3 and P_4 could contribute.

V_1: $\alpha = a$, $\beta = 0$ and $\gamma = n - a$. Then, P_3 and P_4 are of type $r(a,n,a)$. P_2 is of type $l(n,0,n)$. 23 and 24 are n-crossings and 34 is a 0-crossing. First, we observe that ∞ cannot be the point 1 because it would be in 34^+, but 34 is a 0-crossing. Therefore, we have to consider only $\infty = 2$.

23, 24 and 34 could contribute together to $W_{2k-1}(hm = 23)(P_3)$ and they cannot contribute to $-W_{2k-1}(hm = 23)(\bar{P}_3)$ because the foot of 24 would be before the foot of $hm = 23$. But then they contribute also together to $-W_{2k-1}(hm = 24)(\bar{P}_4)$ and they cannot contribute to $W_{2k-1}(hm = 24)(P_4)$ because the foot of 23 would now be before the foot of $hm = 24$. Consequently, their contributions cancel out together. We use here the fact that 23 and 24 have a common foot in P_2 and hence an invisible n-crossings can have its foot in 23^+ and in 24^+ only simultaneously.

We have now to consider the configurations of hm with exactly one other persistent crossing of the tetrahedron.

P_3: contributes $X(23, 34) + II(23, 24)$
$-\bar{P}_3$: contributes $-II(23, 34)$
P_4: contributes $II(24, 34)$
$-\bar{P}_4$: contributes $-X(23, 24) - X(24, 34)$.

But it follows from the triple crossing $P_2 = \bar{P}_2 = 234$ that $X(23, 34) = X(23, 24) + X(24, 34)$ and that $II(23, 34) = II(23, 24) + II(24, 34)$. Consequently, $R^{(2k)}_{a,d^+,hm} = 0$ on the meridian of the quadruple crossing in this case.

V_2: $\alpha = 0$, $\beta = a$ and $\gamma = n - a$. Then, P_4 is of type $r(a, n, a)$ and $hm = 24$ is the only persistent crossing in the tetrahedron. Then, $R^{(2k)}_{a,d^+,hm} = 0$ on the meridian of the quadruple crossing automatically.

V_3: $\alpha = n$, $\beta = 0$ and $\gamma = 0$. Then, P_3 and P_4 are of type $r(n, n, n)$ and there is no stratum of type $r(a, n, a)$ at all.

V_4: $\alpha + \beta = n$ and hence $\gamma = 0$. Then, P_4 is of type $r(n, n, n)$ and again there is no stratum of type $r(a, n, a)$.

Global type VI
All strata are of type l and there is no contribution at all.
We have proven that $R^{(2k)}_{a,d^+,hm}$ satisfies the positive global tetrahedron equation (b). The proof for $R^{(2k)}_{a,ml^-,hm}$ is completely analogous and is left to the reader.

Remark 3.1 *It seems to us that the proof indicates that our method rather replaces analysis than algebra. The point in analysis is that local conditions can lead to a global result, e.g., any smooth function on a compact closed manifold has a critical point.*

The persisting intersection property is a local condition (which holds for each positive triple crossing). We use it to show that sometimes there is an invisible first n-crossing (i.e., not contained in the tetrahedron), which is a global result.

The case of $R_{a,d^+}^{(2k)}$.

The proof follows the same lines as for $R_{a,d^+,hm}^{(2k)}$, but it is more complicated because of the possible n-crossings 23, 24 and 34 for which the foot could slide over the foot or the head of hm from P_i with respect to $\bar{P}_i$. If such a crossing is the first n-crossing, then it contributes only if its foot is in hm^+. Note that by definition hm will now never be the first n-crossing in $W_{2k}(P_i)$, but it could enter as an ordinary n-crossing into a configuration which contributes to $R_{a,d^+}^{(2k)}$. An additional difficulty is that a new type, namely, $r(n,n,n)$, will now also contribute.

Again, in each of the six global types, we have to consider how $W_{2k}(p)$ changes and how this changing is compensated by a configuration with $W_{2k-1}(hm)$ for another stratum. But we have to also study the possible changing of $W_{2k-1}(hm)$ from P_i to $\bar{P}_i$ because of different positions of 0-crossings and n-crossings in the weights. We do all this by inspecting simultaneously a figure from Fig. 3.5 to Fig. 3.16 and the corresponding figure from Fig. 3.17 to Fig. 3.28. We are conscious that this part of the proof will give a hard time to the reader. We apologize for that.

Using Figs. 3.17–3.28, we go again through the list of all cases where at least one of the strata P_i is of the global type $r(a,n,a)$ or $r(n,n,n)$. All other global types of triple crossings will not contribute to $R_{a,d^+}^{(2k)}$.

Global type I

I_1: $\alpha = a$, $\beta = n - a$ and hence $\gamma = 0$. Then, P_1, P_3 and P_4 are of type $r(a,n,a)$ and P_2 is of type $r(n,n,n)$. In this case, 23, 24 and 34 are all n-crossings. $l(P_1) = l(\bar{P}_1) + 1$. There aren't any $(n-a)$-crossings and hence $-l(P_2) + l(\bar{P}_2) = 0$.

23, 24 and 34 contribute together (i.e., we always assume in this case that one of them is the first n-crossing because if there is an invisible first n-crossing, then the contributions cancel out automatically) to $-W_{2k}(P_1)$ if and only if they contribute together also to $W_{2k}(\bar{P}_1)$. If 23, 24 and 34 contribute together to $-W_{2k}(\bar{P}_3)$, then they do not contribute together to $W_{2k}(P_3)$ because now $hm = 23$ would be the first n-crossing. But in this case, they contribute together also in $W_{2k}(P_4)$ and not in $-W_{2k}(\bar{P}_4)$. Consequently, their contributions cancel out in the meridian.

We consider now the case that exactly one, say c, of the three crossings 23, 24 and 34 contributes (i.e., it is the first n-crossing and none of the other two crossings contributes with it together). We denote its contribution by $W_{2k-1}(c)$.

Note that in the case of P_3, we could have $\infty = 2$. But then neither 24 nor 34 could be the first n-crossing because from $\infty = 2$ we would always arrive first to a head of 24 or 34.

Consequently, we have to only study the case $\infty = 1$. Note also that $l(P_3)$ is completely determined by intersections of n-crossings with the crossing $ml = 12$. But this is also the case for P_4 which has the same $ml = 12$. It follows that $l(P_3) = l(\bar{P}_4)$.

We give the contributions for each of the six remaining strata.

$-P_1$: $-(W_{2k-1}(23) + W_{2k-1}(24) + l(P_1)W_{2k-1}(34))$

$\bar{P}_1$: $W_{2k-1}(23) + W_{2k-1}(24) + (l(P_1) - 1)W_{2k-1}(34)$

P_3: $W_{2k-1}(34) + l(P_3)W_{2k-1}(23)$

$-\bar{P}_3$: $-(W_{2k-1}(24) + (l(P_3) + 1)W_{2k-1}(23))$

P_4: $W_{2k-1}(23) + (l(P_3) + 1)W_{2k-1}(24)$

$-\bar{P}_4$: $-l(P_3)W_{2k-1}(24)$.

One easily verifies that these contributions cancel out together on the meridian.

It remains to consider the most complicated case, namely, that exactly two of the three crossings 23, 24 and 34 contribute together. The corresponding notations were already introduced in Definition 3.1.

We give the contributions for each of the six remaining strata.

$-P_1$: $-(X(23, 24) + II(24, 34) + II(23, 24)$

$\bar{P}_1$: $X(23, 24) + II(23, 34) + X(24, 34)$

P_3: $l(P_3)(X(23,24) + II(23,34))$

$-\bar{P}_3$: $-(II(24,23) + II(24,34) + (l(P_3)+1)X(23,34))$

P_4: $X(23,24) + II(23,34) + (l(P_3)+1)X(24,34)$

$-\bar{P}_4$: $-l(P_3)(II(24,34) + II(24,23))$.

As usual, the stratum $P_2 = \bar{P}_2 = 234$ gives the relations $X(23,34) = X(23,24) + X(24,34)$ and $II(23,34) = II(23,24) + II(24,34)$. One easily verifies now that the total contribution to $R_{a,d^+}^{(2k)}$ in this case also vanishes on the meridian.

I_2: $\alpha = a$, $\beta = 0$ and $\gamma = n - a$. Then, P_1 and P_2 are of type $r(a,n,a)$ and P_3 and P_4 are of type $r(a,a,n)$. Only 12 is an n-crossing, but its position with respect to $hm = 34$ does never change because 12 and 34 cannot be together in a triple crossing for the tetrahedron. There aren't any f-crossings (i.e., first n-crossings with the foot in hm^+) which could change because they can appear only for P_3 and P_4. $-l(P_1) + l(\bar{P}_1) = 0$ and $-l(P_2) + l(\bar{P}_2) = 0$.

It follows that $R_{a,d^+}^{(2k)} = 0$ on the meridian of the quadruple crossing in this case.

I_3: $\alpha = 0$, $\beta = a$ and $\gamma = n - a$. Then, P_2 is of type $r(a,n,a)$ and $-l(P_2) + l(\bar{P}_2) = 0$. P_1 is of type $r(0,n,0)$. There are no other n-crossings and hence $hm = 34$ stays the first n-crossing in both $-P_2$ and $\bar{P}_2$. The two 0-crossings 13 and 14 together with 34 form the triple crossing $P_1 = \bar{P}_1 = 134$. Consequently, $-W_{2k-1}(hm = 34)(P_2) + W_{2k-1}(hm = 34)(\bar{P}_2) = 0$ and $R_{a,d^+}^{(2k)} = 0$ on the meridian.

I_4: $\alpha = n - a$, $\beta = a$ and hence $\gamma = 0$. Then, P_2 is of type $r(n,n,n)$. $-l(P_2) + l(\bar{P}_2) = 0$ and $R_{a,d^+}^{(2k)} = 0$ on the meridian.

I_5: $\alpha = n$ and hence $\beta = 0$ and $\gamma = 0$. Then, all four P_i are of type $r(n,n,n)$. There aren't any $(n-a)$-crossings and hence $l(P_i) - l(\bar{P}_i) = 0$ for all four i, and of course $l(P_3) = l(P_4)$. Remember that the crossings ml and d can never contribute to $W_{2k-1}(hm)$ for the global type $r(n,n,n)$. All crossings in the tetrahedron are n-crossings. For P_1, $\bar{P}_1$, P_2 and $\bar{P}_2$, there is no

n-crossing at all which could contribute to $W_{2k-1}(hm)$. Again, we have to only consider the case $\infty = 1$. Exactly as in the case I_1 for $R^{(2k)}_{a,d^+,hm}$, 23, 24 and 34 could contribute together for P_3 and they would no longer contribute together in $-\bar{P}_3$ because of the foot of 24. However, they could contribute together also in $-\bar{P}_4$, and they could not in P_4 because of the foot of 23. But their configurations in P_3 and in $-\bar{P}_4$ correspond exactly to the triple crossing $P_2 = \bar{P}_2 = 234$ before and after the R III move. Moreover, $l(P_3) = l(P_4)$. Consequently, by the preserving intersection property (compare the Introduction), their contributions to $R^{(2k)}_{a,d^+}$ cancel out.

If only the crossing hm enters into the configurations, then of course the contributions of P_i and $\bar{P}_i$ cancel out together.

It remains to consider the case when hm and exactly one other n-crossing enters the configurations.

In this case, the contributions are as follows:

P_3: $l(P_3)(X(23,24) + II(23,34))$
$-\bar{P}_3$: $-l(P_3)X(23,34)$
P_4: $l(P_3)X(24,23)$
$-\bar{P}_4$: $-l(P_3)(II(24,23) + II(24,34))$.

Again, by the Preserving Intersection property for P_2, the contributions cancel out together on the meridian.

I_6: $\alpha = 0$, $\beta = n$ and hence $\gamma = 0$. Then, P_2 is of type $r(n,n,n)$ and $-l(P_2) + l(\bar{P}_2) = 0$. None of the 0-crossings 12, 13 and 14 could contribute together with $hm = 34$ in P_2 or $\bar{P}_2$ for $\infty = 1$. Therefore, we have to only consider the case $\infty = 4$. Here, a new and very interesting phenomena appears again.

The first foot property allows us for certain global types of triple crossings a more precise application of the preserving intersection property because of the condition that a certain crossing has to be the first n-crossing.

$-P_2$: $-X(34,14)$
$\bar{P}_2$: $X(34,13) + II(34,14)$.

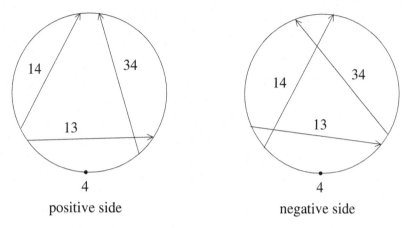

positive side   negative side

Figure 3.32:   The two sides of P_1 of type $r(0, n, 0)$.

Hence, *a priori* the contributions wouldn't cancel out. However, the crossings 14, 13 and 34 form together the triple crossing $P_1 = \bar{P}_1 = 134$, which is of type $r(0, n, 0)$. We show both sides of this stratum in Fig. 3.32.

In $II(34, 14)$, on the positive side, 34 has to be the first n-crossing. However, it follows from the preserving intersection property that $II(34, 14)$ corresponds to $II(34, 13)$ on the negative side. But this is not possible because 34 is no longer the first n-crossing and hence there is an invisible first n-crossing. But this implies that in fact $II(34, 14) = 0$ because 34 can never be the first n-crossing. Moreover, $X(34, 13)$ on the positive side corresponds either to $X(34, 14)$ or to $X(14, 13)$ on the negative side. But again, $X(14, 13)$ is only possible if there is an invisible first n-crossing. In this case, 34 wasn't the first n-crossing in $X(34, 13)$ on the positive side. It follows that $X(34, 13) = X(34, 14)$ and consequently $R_{a,d^+}^{(2k)} = 0$ on the meridian.

I_7: $\alpha = $ arbitrary, $\beta = 0$ and $\gamma = 0$. Then, P_3 is of type $r(n, n, n)$. If $\alpha = n - a$, then $[24] = [14]$, and they contribute simultaneously to $l(p)$. Consequently, $l(P_3) - l(\bar{P}_3) = 0$ in any case. Again, we need to consider only $\infty = 1.14$ can never contribute with $hm = 23$ as the first n-crossing. $[24] = [34] = \alpha$ and we have to consider the cases $\alpha = 0$ and $\alpha = n$. If $\alpha = 0$, then $\infty = 1$ is in 24^+ for P_2, which is not possible because $[24] = 0$. It remains to consider the case $\alpha = n$, where all strata are of type $r(n, n, n)$.

This time $\infty = 4$ is not possible for P_2 because it is in 12^- with the homological marking $[12] = n$.

For P_1 and $\bar{P}_1$, none of the three crossings could contribute with $hm = 34$ as the first n-crossing. The same is true for P_2 and $\bar{P}_2$ because $\infty = 1$. P_3 and P_4 share the same crossing $ml = 12$ and there are no $(n - a)$-crossings in the tetrahedron. It follows that $l(P_3) = l(\bar{P}_3) = l(P_4) = l(\bar{P}_4)$ and we can ignore this global factor. The only non-trivial contributions now are

P_3: $X(23, 24) + II(23, 34)$

$-\bar{P}_3$: $-X(23, 34)$

P_4: $X(24, 34)$

$-\bar{P}_4$: $-(II(24, 23) + II(24, 34))$.

The crossings 23, 24 and 34 form the triple crossing P_2 and hence we have the usual relations from the preserving intersection property. (Of course, we use all the time that 23 and 24 have the same foot, and hence they can be the first n-crossing with other crossings only simultaneously.)

The three crossings can contribute together exactly in P_3 and in $-\bar{P}_4$, and their contributions cancel out. It follows that $R_{a,d^+}^{(2k)} = 0$ on the meridian.

Global type II

Here, only the strata P_1 and P_4 could contribute.

II_1: $\alpha = a$, $\beta = 0$ and $\gamma = n - a$. Then, P_1 and P_4 are of type $r(a, n, a)$. $-l(P_1) + l(\bar{P}_1) = 0$ because $[12] = a$, and $l(P_4) - l(\bar{P}_4) = 0$. We have $[23] = 0$ and $[24] = [34] = n$. The crossings $hm = 34$, 23 and 24 cannot contribute together in $-P_1$. Indeed, they would have to contribute together also in $\bar{P}_1$, but the head of 23 shows that there has to be an invisible n-crossing. The same is true for P_4 and $-\bar{P}_4$. In the latter case, $hm = 24$ can even not contribute with neither 23 or 24. However, we have the f-crossing 34. Consequently, the contributions are

$-P_1$: $-(X(34, 24) + II(34, 23))$

$\bar{P}_1$: $II(34, 24)$

P_4: $II(34, 24)$

$-\bar{P}_4$: $-(X(34, 24) + II(34, 23))$.

The three crossings form together the triple crossing $P_2 = \bar{P}_2 = 234$, which is of type $l(n, n, 0)$. Considerations completely analogous to those in I_6 show now that

$$X(34, 24) = 0 \quad \text{and} \quad II(34, 23) = II(34, 24).$$

It follows that $R_{a,d^+}^{(2k)} = 0$ on the meridian.

II_2: $\alpha = a$, $\beta = n - a$ and hence $\gamma = 0$. Then, P_4 is of type $r(a, n, a)$. [23] $= n - a$ and f-crossings do not change. $l(P_4) - l(\bar{P}_4) = 0$ and again $hm = 24$ cannot contribute with the n-crossing 34. Hence, $R_{a,d^+}^{(2k)} = 0$ on the meridian.

II_3: $\alpha = n$, $\beta = 0$ and $\gamma = 0$. Then, P_1 and P_4 are of type $r(n, n, n)$. P_2 is of type $l(n, n, 0)$. There are no $(n - a)$-crossings at all and hence $l(p)$ can never change. For P_4 and $-\bar{P}_4$, none of the crossings 23 and 34 could contribute together with $hm = 24$. For $-P_1$, the three crossings cannot contribute together again because the head of 23 in $\bar{P}_1$ shows that there is an invisible first n-crossing. Consequently, the only contributions are

$-P_1$: $-(X(34, 24) + II(34, 23))$

$\bar{P}_1$: $II(34, 24)$.

Again, considerations completely analogous to those in I_6 now show that

$$X(34, 24) = 0 \quad \text{and} \quad II(34, 23) = II(34, 24).$$

Hence, $R_{a,d^+}^{(2k)} = 0$ on the meridian.

Global type III
Here, only the strata P_1 and P_2 could contribute. Consequently, f-crossings cannot change.

III_1: $\alpha = a$, $\beta = 0$ and $\gamma = n - a$. Then, P_1 is of type $r(a, n, a)$. [12] $= a$ and hence $-l(P_1) + l(\bar{P}_1) = 0$. For $\infty = 1$, none of the three crossings could contribute with $hm = 34$. For $\infty = 3$, we have

$-P_1$: $-(X(34,23) + II(34,24)$

$\bar{P}_1$: $X(34,24)$.

The three crossings form the triple crossing $P_2 = \bar{P}_2 = 234$ and considerations completely analogous to those in I_6 now show that $II(34,24) = 0$ and that $X(34,24) = X(34,23)$.
Hence, $R_{a,d+}^{(2k)} = 0$ on the meridian.

III_2: $\alpha = 0$, $\beta = a$ and $\gamma = n - a$. Then, P_1 and P_2 are of type $r(a, n, a)$. $[12] = 0$, $[13] = a$ and hence $-l(P_1) + l(\bar{P}_1) = 0$ and $-l(P_2) + l(\bar{P}_2) = 0$. For $\infty = 1$, none of the three crossings could contribute with hm neither for P_1 nor for P_2. For $\infty = 3$, the crossing 12 can still not contribute neither with $-P_1$ nor with $\bar{P}_1$.

III_3: $\alpha = n - a$, $\beta = a$ and $\gamma = 0$. Then, P_1 is of type $r(n, n, n)$ and P_2 is of type $r(a, n, a)$. Here, an unexpected phenomenon arrives: $[13] = n$ and it follows that $-l(P_2) + l(\bar{P}_2) = 1$ (this is the *only* case in the global tetrahedron equation where this happens). But 34 is also the crossing hm in P_1, which is of type $r(n, n, n)$. $[12] = n - a$ and hence $-l(P_1) + l(\bar{P}_1) = 1$ too, but only for $\infty = 1$, because $\infty = 3$ does not have the right position with respect to 12 (compare Definition 2.4). None of the three crossings contributes to $W_{2k-1}(hm)$ neither for P_1, $\bar{P}_1$ nor for P_2, $\bar{P}_2$. The n-crossings 13 and 14 are both f-crossings for $-P_2$ and $\bar{P}_2$. Remember that the contributions of the strata $r(n, n, n)$ enter into the formula with a negative global sign, and hence we obtain the following contributions:

$-P_1 + \bar{P}_1$: $-W_{2k-1}(hm = 34)(P_1)$

$-P_2 + \bar{P}_2$: $W_{2k-1}(hm = 34)(P_2) - X(14,34) - X(13,14) - II(13,34) + X(13,34) + II(14,34) + II(14,13)$.

The three crossings form together the triple crossing $P_1 = \bar{P}_1 = 134$. The three crossings could also contribute together to $-W_{2k}(P_2)$ but then they also contribute together to $W_{2k}(\bar{P}_2)$ and cancel out. Evidently, $W_{2k-1}(hm = 34)(P_1) = W_{2k-1}(hm = 34)(P_2)$ because the crossings 13 and 14 can never contribute. It remains to study the contributions from the f-crossings. We show the two sides of $P_1 = r(n, n, n)$ in Fig. 3.33.

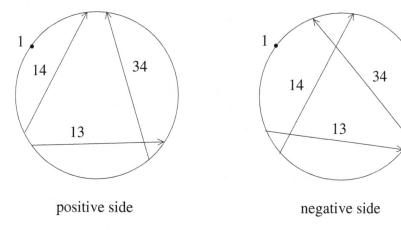

positive side   negative side

Figure 3.33:   The two sides of P_1 of type $r(n,n,n)$.

The preserving intersection property in this case with the first crossing in the couple (c, c') gives the following first n-crossing:

$$X(13, 34) = X(13, 14) + X(14, 34) \quad \text{and}$$
$$II(13, 34) = II(14, 13) + II(14, 34).$$

It follows that $R_{a,d^+}^{(2k)} = 0$ on the meridian.

This is the only case in the global tetrahedron equation where the contributions of strata $r(a, n, a)$ are compensated by the contributions of strata $r(n, n, n)$!

III_4: $\alpha = a$, $\beta = n - a$ and $\gamma = 0$. Then, P_1 is of type $r(n, n, n)$. There aren't any other persistent crossings in the tetrahedron. $[12] = a$ and we have to distinguish two cases. If $a \neq n - a$, then $-l(P_1) + l(\bar{P}_1) = 0$ and the contributions of $-P_1$ and $\bar{P}_1$ cancel out together. If $a = n - a$ and $\infty = 1$, then $-l(P_1) + l(\bar{P}_1) = 1$ and they contribute $-W_{2k-1}(hm = 34)(P_1)$. But exactly in this case P_2 is of type $r(n - a = a, n, n - a = a)$ with $\infty \in d^+$. $[13] = n$ and $-l(P_2) + l(\bar{P}_2) = 1$. Hence, $-P_2 + \bar{P}_2$ contribute $W_{2k-1}(hm = 34)(P_2) - X(14, 34) - X(13, 14) - II(13, 34) + X(13, 34) + II(14, 34) + II(14, 13)$ exactly as was considered before in case III_3.

It follows that $R_{a,d^+}^{(2k)} = 0$ on the meridian.

Global type IV
Here, only the strata P_2 and P_3 could contribute.

IV_1: $\alpha = 0$, $\beta = a$ and $\gamma = 0$. Then, P_2 is of type $r(a, n, a)$ and there aren't any f-crossings. $[14] = n$, $[13] = 0$ and $-l(P_2) + l(\bar{P}_2) = 0$. The contributions are as follows:

$-P_2$: $-II(34, 14)$
$\bar{P}_2$: $II(34, 13) + X(34, 14)$.

The three crossings form $P_1 = \bar{P}_1 = 134$ which is of type $l(n, n, 0)$.
Again, considerations completely analogous to those in I_6 now show that

$$X(34, 14) = 0 \quad \text{and} \quad II(34, 13) = II(34, 14).$$

Hence, $R_{a,d+}^{(2k)} = 0$ on the meridian.

IV_2: $\alpha = a$, $\beta = n - a$ and $\gamma = 0$. Then, P_3 is of type $r(a, n, a)$. $[34] = [24] = n - a$ and f-crossings cannot change. $l(P_3) - l(\bar{P}_3) = 0$ and $[14] = n$. But 14 and 23 cannot be together in a triple crossing, and hence, even for $\infty = 2$, the contribution of 14 to $W_{2k-1}(hm = 23)(P_3)$ and to $W_{2k-1}(hm = 23)(\bar{P}_3)$ is the same and they cancel out together.

IV_3: $\alpha = a$, $\beta = 0$ and $\gamma = n - a$. Then, P_3 is of type $r(a, n, a)$ and $\infty = 2$. $[34] = [24] = 0$ and f-crossings cannot change. $l(P_3) - l(\bar{P}_3) = 0$ and none of the three crossings can contribute with $hm = 23$.

Hence, $R_{a,d+}^{(2k)} = 0$ on the meridian.

IV_4: $\alpha = 0$, $\beta = n$ and $\gamma = 0$. Then, P_2 is of type $r(n, n, n)$. $-l(P_2) + l(\bar{P}_2) = 0$ and $\infty = 1$, $[13] = 0$ and $[14] = n$. The contributions are

$-P_2$: $-II(34, 14)$
$\bar{P}_2$: $X(34, 14) + II(34, 13)$.

The three crossings form $P_1 = \bar{P}_1 = 134$ which is of type $l(n, n, 0)$.

Again, considerations completely analogous to those in I_6 now show that

$$X(34, 14) = 0 \quad \text{and} \quad II(34, 13) = II(34, 14).$$

Hence, $R_{a,d+}^{(2k)} = 0$ on the meridian.

Global type V

Here, only the strata P_3 and P_4 could contribute.

V_1: $\alpha = a$, $\beta = 0$ and $\gamma = n - a$. Then, P_3 and P_4 are of type $r(a, n, a)$. P_2 is of type $l(n, 0, n)$. $[34] = 0$, $[23] = [24] = n$ and hence both f-crossings change. $\infty = 1$ is not possible because $1 \in 34^+$, which is in contradiction with $[34] = 0$. We have to only consider $\infty = 2$. Evidently, $l(P_3) = l(\bar{P}_4)$ and we observe that if 24, 34 and $hm = 23$ contribute together in $W_{2k-1}(hm = 23)(P_3)$, then they also contribute together in $-W_{2k-1}(hm = 24)(\bar{P}_4)$ and cancel out because the three crossings form together the stratum $P_2 = \bar{P}_2 = 234$. Let c be an n-crossing of the tetrahedron. We introduce the notation $W(c)$ for $W_{2k-1}(c)$, i.e., c is the first n-crossing but *without* taking into account in the configurations any other persistent crossing from the tetrahedron (in other words, we only consider c together with the invisible persistent crossings). We write now the contribution of each of the four strata:

P_3: $l(P_3)(W(23) + X(23, 34) + II(23, 24))$

$-\bar{P}_3$: $-(W(24) + X(24, 23) + X(24, 34) + (l(P_3) + 1)(W(23) + II(23, 34)))$

P_4: $W(23) + X(23, 34) + II(23, 24) + (l(P_3) + 1)(W(24) + II(24, 34))$

$-\bar{P}_4$: $-l(P_3)(W(24) + X(24, 23) + X(24, 34))$.

It follows that the total contribution of the four strata is

$$(l(P_3) + 1)(X(23, 34) - X(24, 23) - X(24, 34))$$
$$-(l(P_3) + 1)(II(23, 34) - II(23, 24) - II(24, 34)).$$

The stratum $P_2 = \bar{P}_2 = 234$ is of type $l(n, 0, n)$ and we show the two sides in Fig. 3.34.

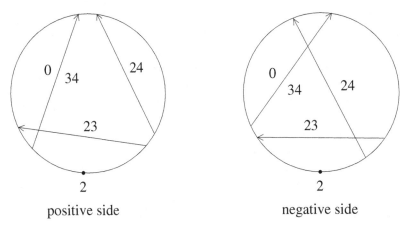

positive side   negative side

Figure 3.34:   The two sides of P_2 of type $l(n, 0, n)$ for the global type V.

The preserving intersection property in this case with the first crossing in the couple (c, c') gives the following first n-crossing:

$$X(23, 34) = X(24, 23) + X(24, 34) \quad \text{and}$$

$$II(23, 34) = II(23, 24) + II(24, 34).$$

Hence, $R^{(2k)}_{a,d^+} = 0$ on the meridian. (We are really lucky here!)

V_2: $\alpha = 0$, $\beta = a$ and $\gamma = n - a$. Then, P_4 is of type $r(a, n, a)$ and $[23] = n - a$, $[34] = a$, $[13] = 0$. Consequently, $l(P_4) - l(\bar{P}_4) = 0$. Moreover, there aren't any f-crossings at all and no crossing contributes to $W_{2k-1}(hm = 24)(P_4)$ or $W_{2k-1}(hm = 24)(\bar{P}_4)$.

Hence, $R^{(2k)}_{a,d^+} = 0$ on the meridian.

V_3: $\alpha = n$, $\beta = 0$ and $\gamma = 0$. Then, P_3 and P_4 are of type $r(n, n, n)$. P_1 and P_2 are both of type $l(n, 0, n)$. It follows that $l(P_3) = l(\bar{P}_3) = l(P_4) = l(\bar{P}_4)$. $\infty = 1$ is not possible because $1 \in 34^+$, but $[34] = 0$. For $\infty = 2$, we have the following contributions:

P_3: $X(23, 34) + II(23, 24)$

$-\bar{P}_3$: $-II(23, 34)$

P_4: $II(24, 34)$

$-\bar{P}_4$: $-(X(24, 23) + X(24, 34))$.

The stratum $P_2 = \bar{P}_2 = 234$ is of type $l(n, 0, n)$ and we have already shown the two sides in Fig. 3.34. The preserving intersection property in this case with the first crossing in the couple (c, c') gives the following first n-crossing:

$$X(23, 34) = X(24, 23) + X(24, 34) \quad \text{and}$$
$$II(23, 34) = II(23, 24) + II(24, 34).$$

Hence, $R^{(2k)}_{a,d^+} = 0$ on the meridian.

V_4: $\alpha + \beta = n$ and hence $\gamma = 0$. Then, P_4 is of type $r(n, n, n)$. We can assume that $\beta \neq 0$ because this was the case V_3. There are no other persistent crossings because $[34] = \beta$ and $[23] = \alpha + \gamma$. We have $[23] = [13]$ because $\gamma = 0$. It follows that $l(P_4) - l(\bar{P}_4) = 0$ and hence $R^{(2k)}_{a,d^+} = 0$ on the meridian.

Global type VI
All strata are of type l and there is no contribution at all.

We have proven that our 1-cochain $R^{(2k)}_{a,d^+}$ satisfies (b): the positive global tetrahedron equation. This was the hardest part!

The proof for $R^{(2k)}_{a,ml^-}$ is completely analogous and is left to the reader. We only indicate why $l_{ml^-}(p)$ is different from $l(p)$ for the global type $r(n, n, n)$ (remember that the arc of the point ∞ is determined for a persistent crossing, but that this is not the case for non-persistent crossings):

III_3: $\alpha = n-a$, $\beta = a$ and $\gamma = 0$. Then, P_1 is of type $r(n, n, n)$ and P_2 is of type $r(a, n, a)$ with $\infty = 3 \in ml^-$. Here, an unexpected phenomenon arrives: $[13] = n$, and it follows that $-l(P_2)+l(\bar{P}_2) = 1$ (this is the *only* case in the global tetrahedron equation where this happens). But 34 is also the crossing hm in P_1 which is of type $r(n, n, n)$. $[12] = n - a$ and hence $-l_{ml^-}(P_1) + l_{ml^-}(\bar{P}_1) = 1$ too, but only for $\infty = 3$, because $\infty = 1$ now does not have the right position with respect to 12 (compare Definition 2.4). The rest is analogous to the case of $R^{(2k)}_{a,d^+}$.

Remark 3.2 $R^{(2k)}_{a,d^+}$ *is no longer defined for $a = 0$ because ∞ now has to be in ml^-. Moreover, it is clear that the proof breaks down* for $a = 0$ because in $r(a, n, a)$ the crossings ml and d could now contribute together with the n-crossing hm, and the contributions of the two strata P_i and $\bar{P}_i$ enter sometimes with a different refined linking number $l_{ml^-}(p)$.

It also breaks down for $a = n$ because now $r(a, n, a)$ and $r(n, n, n)$ are of the same type, but they enter with different global signs into the formula!

We haven't studied the possible degenerations of $R^{(2k)}_{a,d^+,hm}$ for $a = 0$ or $a = n$.

$R^{(2k)}_{a,d^+,hm}$ and $R^{(2k)}_{a,d^+}$ satisfy the cube equations (c).

If a loop in M_n passes with an ordinary homotopy through a triple crossing where two branches are ordinary tangential, then in the R III move exactly one crossing is replaced by another crossing with the same homological marking but with the opposite sign. Evidently, we have to only study the case when this crossing is the n-crossing hm for the type $r(a, n, a)$ or $r(n, n, n)$. We show the corresponding edges of the graph Γ (compare Section 3.1) in Figs. 3.35–3.38. For the numbers or names of the local types of triple crossings, compare Fig. 2.2.

<h2 style="text-align:center">r: 1–7</h2>

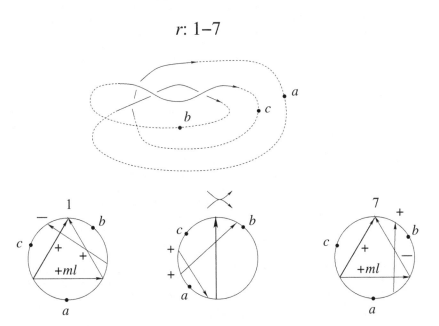

Figure 3.35:   Edge $r1 - 7$.

r: 4–6

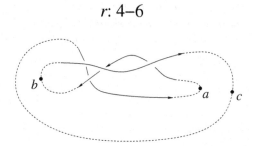

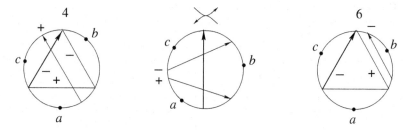

Figure 3.36:   Edge *r*4 − 6.

r: 5–2

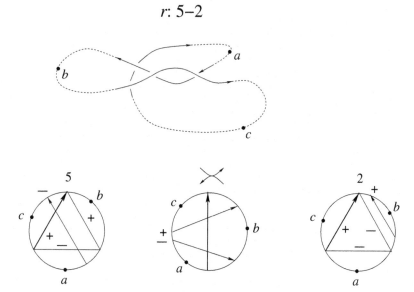

Figure 3.37:   Edge *r*5 − 2.

r: 3–8

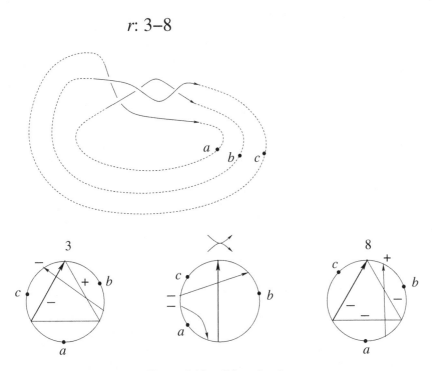

Figure 3.38:   Edge $r3 - 8$.

The positive triple crossing corresponds to type 1. As already mentioned, the global type of two triple crossings of an edge is always the same. For completeness, we show as an example the unfolding of the edge $r1 - 5$ in M_n in Fig. 3.39, see Ref. [16].

It is very convenient that with our definition the crossing hm enters $W_{2k-1}(hm)w(hm)$ always with a positive sign. The other new n-crossing, lets call it hm', from the R II move does never enter together with hm into a configuration. Consequently, $W_{2k-1}(hm)w(hm)$ is the same for the two R III moves in the edge of Γ. This already shows that $R^{(2k)}_{a,d^+,hm}$ satisfies the cube equations (c).

Let us consider the new crossing hm' from the R II move. The configuration with hm' which has its foot in hm^+ now contributes to $W_{2k}(p)$ by $w(hm')W_{2k-1}(hm)$ and the linking number $l(p)$ changes by adding $w(hm')$. Going through Figs. 3.35–3.38, we see that in each case with the correction term $(w(hm) - 1)W_{2k-1}(hm)w(hm)$

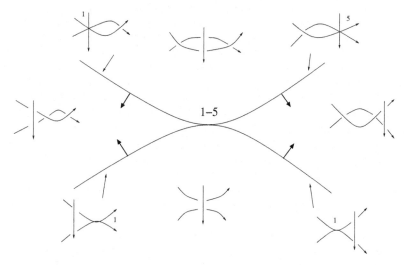

Figure 3.39:   The unfolding of the edge $r1 - 5$.

the contributions of the two R III moves in the edge are now the same.

If the crossings from the R II moves do not correspond to the crossing hm, then the two crossings hm in the R III moves have the same sign and the correction terms for both R III moves are the same.

For the R III moves of type $r(n, n, n)$ (where we do not consider $W_{2k}(p)$ at all) again hm enters $W_{2k-1}(hm)w(hm)$ always with a positive sign. Moreover, the new n-crossing never enters $l(p)$ because only $(n - a)$-crossings can contribute to $l(p)$ for this type (and $0 < a < n$). This shows that $R_{a,d^+}^{(2k)}$ satisfies the cube equations (c) too.

The proofs for $R_{a,ml^-,hm}^{(2k)}$ and $R_{a,ml^-}^{(2k)}$ are exactly the same. $\square$

We have already proven that $R_{a,d^+,hm}^{(2k)}$ is well defined in $\tilde{M}_n$ (i.e., all Reidemeister I moves are allowed as well) in Section 2.2. The same is true of course also for $R_{a,ml^-,hm}^{(2k)}$.

$R_{a,d^+}^{(2k)}$ is well defined in $M_n^{\text{semi-reg}}$, i.e., it vanishes on the meridian of $\Sigma_{\text{trans-cusp}}^{(2)}$ (compare Proposition 1.1) if the new crossing is a 0-crossing. Indeed, in this case, only the new R III move can neither be of type $r(a, n, a)$ nor be of type $r(n, n, n)$. However, if the new crossing from the R I move is an n-crossing, then it could be the crossing hm in an R III move p of type $r(a, n, a)$. The arc ml^- need not to be empty of foots of crossings in this case and could give lots of contributions to $W_{2k}(p)$. Hence, $R^{(2k)}_{a,d+}$ is not well defined in $\tilde{M}_n$. The same is true of course also for $R^{(2k)}_{a,ml-}$.

The examples in Chapter 4 show that all four 1-cocycles represent non-trivial cohomology classes in M_n and in particular they are not always trivial on the loop $rot(\sigma_1 \cup 2K)$, which is a very particular case of the pairing from Theorem 2.1. This finishes the proof of Theorems 2.1–2.4. □

3.3 Proof of Theorem 2.5

We have already proven in the previous section that $\bar{R}^{(2k)}_a$ satisfies the commutation relations (a).

The 1-cochain $\bar{R}^{(2)}_a$ satisfies the positive global tetrahedron equation (b).

The global types of quadruple crossings come in pairs by inverting all arrows (but without changing the sign of the crossings): I with VI, II with IV and III with V. However, our weights based on the configuration $(n, 0)$ are not invariant under this operation because it becomes $(0, n)$, i.e., going from ∞ we come first to the overcross of a 0-crossing. Therefore, we have to check that $\bar{R}^{(2k)}_a$ vanishes on the meridians again for all six global types of positive quadruple crossings.

Note that there is no longer the factor $l(p)$ in the contributions of the global type $r(a, n, a)$. We have to show now (b) that $\bar{R}^{(2)}_a$ without the correction terms from the cube equations $1/2(w(hm) - 1)\bar{W}_1(hm)w(hm)$ (which vanish automatically for positive triple crossings) vanishes on the meridian of each global positive quadruple crossing, i.e., it satisfies the positive global tetrahedron equation. Since all crossings are positive, if we have to chose a side, then it is always the positive side of the triple crossing.

We now have to consider *two* generic types of triple crossings, $r(a, n, a), \infty \in d^+$, and $l(a, 0, a), \infty \in ml^+$, and lots of degenerate types, but only the single configuration C_2. Moreover, one easily sees that the individual contribution of an n-crossing to $C_2 = (n, 0)$

changes only by passing a triple crossing of type $l(n, 0, n)$ and that the individual contribution of a 0-crossing to $C_2 = (n, 0)$ changes only by passing a triple crossing of type $r(0, n, 0)$.

Note that for degenerate tetrahedrons (i.e., all six crossings are persistent crossings), the point at infinity is always uniquely defined and that we see the homological markings directly in Figs. 3.5–3.16.

We have to study not only the n-crossings but also the 0-crossings which change the arc of their foot with respect to hm from the stratum P_i to the stratum $\bar{P}_i$.

Remember that with respect to hm we have to consider only the following crossings (which were considered already for $R_{a,d^+}^{(2k)}$ for the global type $r(a, n, a)$):

P_3: foot of 24 in $hm^-(hm = 23)$, $\bar{P}_3$: foot of 24 in hm^+

P_3: foot of 34 in $hm^+(hm = 23)$, $\bar{P}_3$: foot of 34 in hm^-

P_4: foot of 23 in $hm^+(hm = 24)$, $\bar{P}_4$: foot of 23 in hm^-.

In particular, f-crossings of marking 0 or n never change for P_1, $\bar{P}_1$, P_2 and $\bar{P}_2$.

For the global types of quadruple crossings I up to VI, we now have to consider all the homological parameters α, β, γ such that there is a stratum of type $r(a, n, a)$ or $l(a, 0, a)$ or a degenerate stratum, which contains at least one n-crossing as well as one 0-crossing. Note that for degenerate tetrahedrons all $l_a(p)$ are evidently the same for P_i and $\bar{P}_i$ because there aren't any a-crossings in the tetrahedron.

We denote the contribution to $\bar{W}_2(p)$, respectively, $\bar{W}_1(hm)$, of an n- or a 0-crossing c with $C_2 = (n, 0)$ simply by $W_1(c)$.

We ignore simply those triple crossings which do not enter the formula for $\bar{R}_a^{(2)}$, as well as those degenerate triple crossings which cannot change the weights, i.e., $r(n, n, n)$ and $l(0, 0, 0)$.

The refined linking numbers $l_a(p)$ for the degenerated triple crossings were introduced in Definitions 2.16–2.19. Let us start with a lemma which will be very important in the proof.

Lemma 3.1 *We have $l_a(P_1) = l_a(P_4)$ for each degenerate positive quadruple crossings (i.e., all six crossings are persistent crossings) of the global types II and IV. (Remember that always $0 < a < n$.)*

Proof. For degenerate tetrahedrons, the homological markings are already determined by the point ∞. We show the three possible types for II in Fig. 3.40. (For the remaining possibility, the strata P_1 and P_4 would both be of type $r(n, n, n)$, and they do not contribute at all to the 1-cochain $\bar{R}_a^{(2k)}$.)

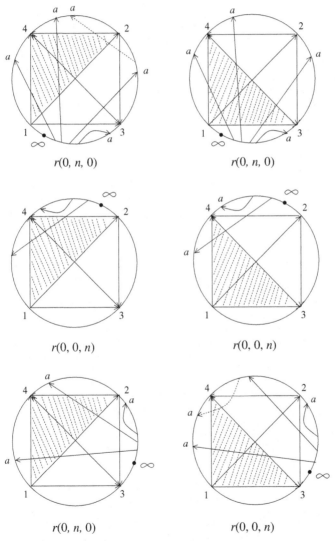

Figure 3.40: $l_a(P_1) = l_a(P_4)$ for the global case II of positive quadruple crossings.

We see that in all cases the contributing a-crossings for $P_1 = 134$ and $P_4 = 124$ are exactly the same. A *priori*, there could be more contributing a-crossings, which we indicate by dashed arrows. However, the dashed arrow together with the n-crossing 34, respectively, with the 0-crossing 12, would create a negative loop in the diagram!

We show the three possible types for IV in Fig. 3.41. (For the remaining possibility, the strata P_1 and P_4 would both be of type $l(0, 0, 0)$, and they do not contribute to the 1-cochain $\bar{R}_a^{(2k)}$ neither.)

We see that again in all cases the contributing a-crossings for $P_1 = 134$ and $P_4 = 124$ are exactly the same. Again, there could be more contributing a-crossings, which we indicate by dashed arrows. However, again the dashed arrow together with the n-crossing 34, respectively, with the 0-crossing 12, would create a negative loop in the diagram, which is not allowed. $\square$

We see immediately from the proof that the lemma is false for $a = 0$ and for $a = n$.

Global type I

All global types of triple crossings in the meridian are of type r in this case.

I_1: $\alpha = a$, $\beta = n - a$ and hence $\gamma = 0$. Then, P_1, P_3 and P_4 are of type $r(a, n, a)$ and P_2 is of type $r(n, n, n)$. In this case, 23, 24 and 34 are all n-crossings. There aren't any 0-crossings and hence all weights $\bar{W}_2$ of crossings stay the same.

$$P_3 - \bar{P}_3 = W_1(34, 23^+) - W_1(24, 23^+)$$
$$P_4 - \bar{P}_4 = W_1(23, 24^+).$$

Hence, the total contribution is $W_1(23, 24^+) + W_1(34, 23^+) - W_1(24, 23^+)$. But 24, 34 and 23 form the triple crossing $P_2 = 234$ and $W_1(24, 23^+) - W_1(34, 23^+) - W_1(23, 24^+) = 0$, as shown in Fig. 3.42. It follows that $\bar{R}_a^{(2)} = 0$ on the meridian.

I_2: $\alpha = a$, $\beta = 0$ and $\gamma = n - a$. Then, P_1 and P_2 are of type $r(a, n, a)$ and P_3 and P_4 are of type $r(a, a, n)$. There aren't any f-crossings which could change because they can appear for the type $r(a, n, a)$ only for P_3 and P_4. There aren't any 0-crossings and all weights stay the same. Hence, $\bar{R}_a^{(2)} = 0$ on the meridian.

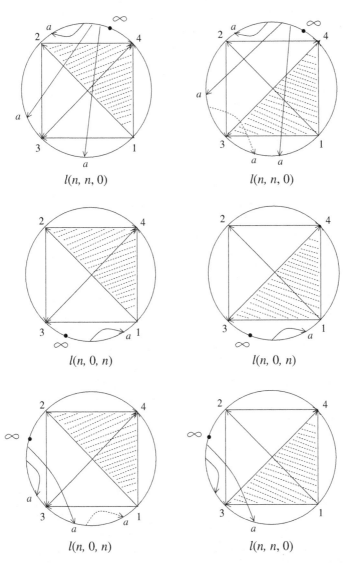

Figure 3.41: $l_a(P_1) = l_a(P_4)$ for the global case IV of positive quadruple crossings.

I_3: $\alpha = 0$, $\beta = a$ and $\gamma = n - a$. Then, P_2 is of type $r(a, n, a)$ and P_1 is of type $r(0, n, 0)$. Only $\infty = 4$ is interesting because $[14] = 0$ and we would have $1 \in 14^+$. We have $[13] = [14] = 0$ and $(34, 14)$ contributes only in $-P_2$ and $(34, 13)$ contributes only

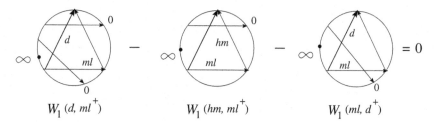

$$W_1(d, ml^+) \qquad\qquad W_1(hm, ml^+) \qquad\qquad W_1(ml, d^+)$$

Figure 3.42: $W_1(24, 23^+) - W_1(34, 23^+) - W_1(23, 24^+) = 0$ for $P_2 = 234$ of type $r(n, n, n)$.

in $\bar{P}_2$. There aren't any persistent crossings in the tetrahedron besides those in the triangle P_1. The crossings 23 and 24 are both a-crossings, but they contribute to both $l_a(P_1)$ and $l_a(\bar{P}_1)$. It follows that $\bar{R}_a^{(2)} = 0$ on the meridian.

I_4: $\alpha = $ arbitrary but $\alpha \neq a$, $\beta = n - a$ and $\gamma = 0$. Then, P_3 is of type $r(a, n, a)$. But then 34 and 24 are not persistent crossings and consequently $\bar{R}_a^{(2)} = 0$ on the meridian.

These were the only non-degenerate cases. We have to consider now the degenerate tetrahedrons. Consequently, for each stratum P_i, the refined linking numbers $l_a(P_i)$ and $l_a(\bar{P}_i)$ are the same. Moreover, a weight W_1 could change exactly if there is a stratum $r(0, n, 0)$ in the tetrahedron.

I_5: $\alpha = 0$, $\beta = n$ and $\gamma = 0$. Then, P_1, P_3 and P_4 are of type $r(0, n, 0)$ and P_2 is of type $r(n, n, n)$ and necessarily $\infty = 4$. In $-P_1$, the n-crossing 24 contributes 0 to $W_1(d, ml^-) - W_1(ml, d^-)$ and the n-crossing 23 contributes 1 to $W_1(ml, d^-)$. In $\bar{P}_1$, the crossing 24 contributes 1 to $W_1(ml, d^-)$. Consequently, the contributions of $-P_1$ and $\bar{P}_1$ cancel out together. In P_3, there is no contribution and the contribution of 24 cancels out again in $-\bar{P}_3$. In P_4, the n-crossing 23 contributes 1 to $W_1(ml, d^-)$. But in $-\bar{P}_4$, the 0-crossing 13 contributes 1 to $W_1(hm, ml^-)$ and the contributions cancel out again.

Hence, $\bar{R}_a^{(2)} = 0$ on the meridian.

I_6: $\alpha = 0$, $\beta = 0$ and $\gamma = n$. Then, P_1 and P_2 are of type $r(0, n, 0)$ and P_3 and P_4 are of type $r(0, 0, n)$, $\infty = 3$. 23 contributes 1 to

$W_1(hm, ml^-)$ for $-P_1$ and 24 contributes 1 to $W_1(hm, ml^-)$ for $\bar{P}_1$, and they cancel out. There are no contributions from the tetrahedron for $-P_2$ and $\bar{P}_2$. 34 contributes 0 to $W_1(d, ml^+) - W_1(hm, ml^+)$ for $-\bar{P}_3$ and it contributes also 0 to $W_1(d, ml^+) - W_1(hm, ml^+)$ for P_4.

It follows that $\bar{R}_a^{(2)} = 0$ on the meridian.

I_7: $\alpha = n$, $\beta = 0$ and $\gamma = 0$. Then, all strata are of type $r(n, n, n)$ and there are no contributions at all.

I_8: $\alpha = 0$, $\beta = 0$ and $\gamma = 0$. Then, P_1, P_2 and P_4 are of type $r(0, 0, n)$ and P_3 is of type $r(n, n, n)$, $\infty = 2$. The weights W_1 cannot change and in P_2 the foots of the crossings cannot change the arc. There aren't any contributions from the tetrahedron for $-P_1$ and $\bar{P}_1$, as well as for P_4 and $\bar{P}_4$.

It follows that $\bar{R}_a^{(2)} = 0$ on the meridian.

Global type II

II_1: $\alpha = a$, $\beta = 0$ and $\gamma = n - a$. Then, P_1 and P_4 are of type $r(a, n, a)$ and P_3 is of type $l(a, 0, a)$, $\infty = 1$. But P_2 is of type $l(n, n, 0)$ and the individual weights cannot change. $[23] = 0$, $[24] = [34] = n$ and hence $P_4 - \bar{P}_4$ contributes $W_1(23, 24^+)$ and $-P_3 + \bar{P}_3$ contributes $W_1(34, 23^-) - W_1(24, 23^-)$.

But P_2 is of type $l(n, n, 0)$. One easily sees that $W_1(ml, d^+)(P_2) = W_1(ml, d^+)(\bar{P}_2) = W_1(23, 24^+)$, and the same is true for $W_1(hm, ml^-) = W_1(34, 24^-)$ and $W_1(d, ml^-) = W_1(24, 23^-)$. One easily sees that $W_1(34, 23^-) = W_1(34, 24^-)$ because 23 and 24 have the same foot in P_2.

The crossings 12, 13 and 14 are a-crossings. 12 and 13 contribute to $l_a(P_2)$ for P_2. But only the crossing 13 contributes to $-l_a(\bar{P}_2)$. Hence, $l_a(P_2) - l_a(\bar{P}_2) = 1$. Consequently, the total contribution is now

$$W_1(23, 24^+) + W_1(34, 23^-) - W_1(24, 23^-) + (W_1(24, 23^-)$$
$$-W_1(34, 23^-) - W_1(23, 24^+)) = 0.$$

It follows that $\bar{R}_a^{(2)} = 0$ on the meridian.

II_2: $\alpha = a$, $\beta = n - a$ and hence $\gamma = 0$. Then, P_4 is of type $r(a, n, a)$. There is no stratum $l(n, 0, n)$ and the individual weights cannot change. $[23] = n - a$ and f-crossings do not change. It follows that $\bar{R}_a^{(2)} = 0$ on the meridian. (This is still the case if we have more general $\beta + \gamma = n - a$ and $\beta \neq 0$.)

II_3: $\alpha = 0$, $\beta = a$ and $\gamma = 0$. Then, P_2 is of type $l(a, 0, a)$ and P_1 is of type $r(0, 0, n)$. The f-crossings never change for P_2 and $\bar{P}_2$. 23 and 24 are a-crossings for $\infty = 2$ and exactly 24 contributes to $l_a(P_1)$ but also to $l_a(\bar{P}_1)$. There are no contributions from crossings of the tetrahedron for $-P_1$ and $\bar{P}_1$ neither for $\infty = 2$ nor for $\infty = 3$.

Consequently, $\bar{R}_a^{(2)} = 0$ on the meridian.
It remains to consider the degenerate tetrahedrons.

II_4: $\alpha = n$, $\beta = 0$ and $\gamma = 0$. Then, P_2 is of type $l(n, n, 0)$ and P_3 is of type $l(n, 0, n)$, $\infty = 1$.

In P_2, the crossing 12 contributes to $W_1(ml = 23, d^+)$, and in $-\bar{P}_2$, the crossing 13 contributes to $W_1(ml = 23, d^+)$, and they cancel out together.

For $-P_3$ and $\bar{P}_3$, none of the crossings of the tetrahedron contributes to any weight.

Consequently, $\bar{R}_a^{(2)} = 0$ on the meridian.

II_5: $\alpha = 0$, $\beta = 0$ and $\gamma = 0$.

Then, P_1 and P_4 are of type $r(0, 0, n)$ and P_3 is of type $l(n, 0, n)$, $\infty = 3$. There are no contributions from the crossings of the tetrahedron for $-P_1$, $\bar{P}_1$, P_4 and $-\bar{P}_4$. The same is true for $-P_3$ and $\bar{P}_3$. Consequently, $\bar{R}_a^{(2)} = 0$ on the meridian.

II_6: $\alpha = 0$, $\beta = n$ and $\gamma = 0$. Then, P_1 is of type $r(0, 0, n)$, P_4 is of type $r(0, n, 0)$, P_2 is of type $l(n, 0, n)$, and P_3 is of type $l(n, n, 0)$, $\infty = 2$. The crossing 23 contributes in $\bar{P}_1$ to $-W_1(hm = 34, ml^+ = 13^+)$. Consequently, the contribution of $-P_1 + \bar{P}_1$ is

$$+l_a(P_1) = l_a(\bar{P}_1).$$

There are no contributions in $-P_2$ and $\bar{P}_2$.

In $-P_3$, the crossing 34 contributes to $-W_1(hm, ml^-)$, and in $\bar{P}_3$, the crossing 24 contributes to $W_1(ml, d^+)$, and the contributions cancel out together.

In $-\bar{P}_4$, the crossing 34 contributes to $-W_1(hm, ml^-)$. Consequently, $P_4 - \bar{P}_4$ contributes

$$-l_a(P_4).$$

But it follows from Lemma 3.1 that $l_a(P_1) = l_a(P_4)$. Consequently, $\bar{R}_a^{(2)} = 0$ on the meridian.

II_7: $\alpha = 0$, $\beta = 0$ and $\gamma = n$. Then, P_1 and P_4 are of type $r(0, n, 0)$ and P_2 is of type $l(n, n, 0)$, $\infty = 4$. The crossings 12 and 24 do not contribute to $-P_1$ because their foots are in ml^+ instead of ml^-. The crossing 12 contributes to $-W_1(hm, ml^-)$ for $\bar{P}_1$. Consequently, the contribution is

$$+l_a(P_1).$$

The contributions to P_2 and $-\bar{P}_2$ cancel out together.

The crossing 34 contributes 0 to $W_1(d, ml^-) - W_1(ml, d^-)$ for P_4. The crossing 34 contributes to $-W_1(ml, d^-)$ for $-\bar{P}_4$. Consequently, the contribution is

$$-l_a(P_4).$$

But again, it follows from Lemma 3.1 that $l_a(P_4) = l_a(P_1)$ and hence $\bar{R}_a^{(2)} = 0$ on the meridian.

Global type III

III_1: $\alpha = a$, $\beta = 0$ and $\gamma = n - a$. Then, P_1 is of type $r(a, n, a)$, P_2 is of type $r(0, n, 0)$, and P_3 and P_4 are of type $l(a, 0, a)$, $\infty = 3$. We have $[12] = [13] = [14] = a$, $[23] = [24] = 0$ and $[34] = n$. The crossing 23 contributes to $W_1(hm = 34)$ for $-P_1$ and the crossing 24 contributes to $W_1(hm = 34)$ for $\bar{P}_1$ and they cancel out.

The crossing 24 contributes $-W_1(24, hm^- = 23^-)$ for $-P_3$. The crossing 34 contributes $W_1(34, hm^- = 23^-)$ to $\bar{P}_3$. The crossing 23 contributes $W_1(23, hm^- = 24^-) + 1$ to $\bar{P}_4$. The 1 comes from the couple $(34, 23)$. The a-crossings are 12, 13, and 14. We have

$l_a(P_2) = l_a(\bar{P}_2) + 1$ because 12 does no longer contribute for $\bar{P}_2$. It follows that the total contribution is

$$-W_1(24, hm^- = 23^-) + W_1(34, hm^- = 23^-) + W_1(23, hm^- = 24^-)$$
$$+ 1 + (W_1(24, 23^-) - W_1(34, 23^-) - (W_1(23, 24^-) + 1)) = 0.$$

The 1 in $W_1(ml = 23, hm^- = 24^-)$ comes from the couple $(hm = 34, ml = 23)$, which contributes on the *positive* side of the move. Remember that $(hm = 34, ml = 23)$ does not contribute to $W_1(hm = 34, ml^- = 23^-)$ because by definition a crossing c is never in c^+ or c^-.

It follows that $\bar{R}_a^{(2)} = 0$ on the meridian.

III_2: $\alpha = 0$, $\beta = a$ and $\gamma = n - a$. Then, P_1 and P_2 are of type $r(a, n, a)$. Only 12 and 34 are persistent crossings and they never intersect here. The f-crossings never change with respect to hm for P_1 and P_2 and hence $\bar{R}_a^{(2)} = 0$ on the meridian.

III_3: $\alpha = n - a$, $\beta = a$ and $\gamma = 0$. Then, P_2 is of type $r(a, n, a)$. There are no 0-crossings, and f-crossings cannot change.

It follows that $\bar{R}_a^{(2)} = 0$ on the meridian.

III_4: $\alpha = a$, $\beta = 0$ and $\gamma = $ arbitrary, but $\gamma \neq n - a$ (otherwise we have III_1). Then, only P_4 is of type $l(a, 0, a)$, $\infty = 3$. But then $[23] \neq 0$ and $[34] \neq n$ and there is no changing of f-crossings at all.

Consequently, $\bar{R}_a^{(2)} = 0$ on the meridian.
It remains to consider the degenerate tetrahedrons.

III_5: $\alpha = \beta = \gamma = 0$. Then, P_1 and P_2 are of type $r(0, 0, n)$ and P_3 is of type $l(n, n, 0)$, $\infty = 2$. There aren't any crossings in the tetrahedron which contribute to any weight for any of the strata.

Consequently, $\bar{R}_a^{(2)} = 0$ on the meridian.

III_6: $\alpha = n$, $\beta = \gamma = 0$. $\infty = 3$. Then, P_2 is of type $r(0, n, 0)$ and P_3 and P_4 are both of type $l(n, 0, n)$, $\infty = 3$.

$-P_2$: 12 to $W_1(d, ml^-) - W_1(ml, d^-)$, 14 to $-W_1(ml, d^-)$
$\bar{P}_2$: 14 to $W_1(d, ml^-) - W_1(ml, d^-)$, 13 to $-W_1(ml, d^-)$.

They cancel out together.

$-P_3$: 24 to $-W_1(ml, d^+)$
$\bar{P}_3$: 14 to $-W_1(ml, d^+)$.

They cancel out together.

$-P_4$: 0
$\bar{P}_4$: 23 to $W_1(d, ml^+) - W_1(ml, d^+)$.

It follows that $\bar{R}_a^{(2)} = 0$ on the meridian.

III_7: $\alpha = \gamma = 0$, $\beta = n$. Then, P_3 and P_4 are of type $l(n, n, 0)$, $\infty = 1$. There aren't any crossings in the tetrahedron which contribute to any weight and hence $\bar{R}_a^{(2)} = 0$ on the meridian.

III_8: $\alpha = \beta = 0$, $\gamma = n$. Then, P_1 and P_2 are of type $r(0, n, 0)$, $\infty = 4$. There aren't any foots in ml^- for $-P_1$. There is only 12 for $\bar{P}_1$, but it does not contribute.

$-P_2$: 14 to $-W_1(hm, ml^-)$
$\bar{P}_2$: 13 to $-W_1(hm, ml^-)$.

They cancel out together.
It follows that $\bar{R}_a^{(2)} = 0$ on the meridian.

Global type IV

IV_1: $\alpha = 0$, $\beta = a$ and $\gamma = 0$. Then, P_2 is of type $r(a, n, a)$ and P_1 is of type $l(n, n, 0)$, $\infty = 1$. The f-crossings never change for P_2 with respect to $\bar{P}_2$. Exactly, 23 and 24 are a-crossings.

There aren't any persistent crossings in the tetrahedron which contribute to the weights for P_1 and $-\bar{P}_1$. Exactly, the crossing 23 contributes to both $l_a(P_1)$ and $l_a(\bar{P}_1)$ and 24 does not contribute. It follows that $\bar{R}_a^{(2)} = 0$ on the meridian.

IV_2: $\alpha = a$, $\beta = n - a$ and $\gamma = 0$. Then, P_3 is of type $r(a, n, a)$. $[34] = [24] = n-a$ and f-crossings cannot change. Hence, $\bar{R}_a^{(2)} = 0$ on the meridian.

IV_3: $\alpha = a$, $\beta = 0$ and $\gamma = n - a$. Then, P_3 is of type $r(a, n, a)$, P_1 and P_4 are of type $l(a, 0, a)$ and P_2 is of type $r(0, 0, n)$, $\infty = 2$. There are no strata $l(n, 0, n)$, $r(0, n, 0)$ and the individual weights cannot change. We have $[34] = [24] = 0$, $[23] = n$ and the a-crossings are 12, 13 and 14.

The f-crossings do not change for P_1, $-\bar{P}_1$.

$P_3 - \bar{P}_3$: $W_1(34, 23^+) - W_1(24, 23^+)$
$-P_4 + \bar{P}_4$: $W_1(23, 24^-)$.

12 and 14 contribute to $l_a(P_2)$ and only 14 contributes to $l_a(\bar{P}_2)$. Consequently, the total contribution is

$$W_1(34, 23^+) - W_1(24, 23^+) + W_1(23, 24^-) + (W_1(24, 23^+)$$
$$- W_1(34, 23^+) - W_1(23, 24^-)) = 0.$$

It follows that $\bar{R}_a^{(2)} = 0$ on the meridian.

If $\alpha = $ arbitrary but $\alpha \neq a$, then only P_4 is of type $l(a, 0, a)$. But in this case, 23 is not a persistent crossing and no weight changes at all.

It remains to consider the degenerate tetrahedrons.

IV_4: $\alpha = 0$, $\beta = n$ and $\gamma = 0$. Then, P_1 and P_4 are of type $l(n, n, 0)$ and P_3 is of type $r(0, n, 0)$, $\infty = 1$. No crossing from the tetrahedron contributes to the weights for P_3 and $-\bar{P}_3$.

The crossing 23 contributes exactly to $W_1(ml, d^+)$ for P_1. Consequently,

$P_1 - \bar{P}_1$: $l_a(P_1)$.

The crossing 23 is also the only crossing which contributes to $W_1(ml, d^+)$ for $-P_4$. Consequently,

$-P_4 + \bar{P}_4$: $-l_a(P_4)$.

We are in type IV and it follows again from Lemma 3.1 that $l_a(P_1) = l_a(P_4)$ and hence, $\bar{R}_a^{(2)} = 0$ on the meridian.

IV_5: $\alpha = n$, $\beta = 0$ and $\gamma = 0$. P_1 and P_4 are of type $l(n, 0, n)$ and P_2 is of type $r(0, 0, n)$, $\infty = 2$.

The crossing 12 contributes 0 to $W_1(d, ml^+) - W_1(hm, ml^+)$ and the crossing 13 contributes to $-W_1(hm, ml^+)$ for $-P_2$. The crossing 14 contributes 0 to $W_1(d, ml^+) - W_1(hm, ml^+)$ and the crossing 12 contributes to $-W_1(hm, ml^+)$ for $\bar{P}_2$. Hence, the contributions cancel out.

In P_1, only 12 contributes to $-W_1(hm, ml^+)$ and there are no contributions in $-\bar{P}_1$. Consequently,

$$P_1 - \bar{P}_1: -l_a(P_1).$$

In $-P_4$, the crossing 34 contributes to $-W_1(ml, d^+)$. In $\bar{P}_4$, the crossing 34 contributes 0 to $W_1(d, ml^+) - W_1(ml, d^+)$. Consequently,

$$-P_4 + \bar{P}_4: l_a(P_4).$$

Again, we have $l_a(P_1) = l_a(P_4)$ from Lemma 3.1. Consequently, $\bar{R}_a^{(2)} = 0$ on the meridian.

IV_6: $\alpha = \beta = \gamma = 0$. P_1 is of type $l(n, n, 0)$, P_2 is of type $r(0, n, 0)$, P_3 is of type $r(0, 0, n)$ and P_4 is of type $l(n, 0, n)$, $\infty = 3$. 23 contributes to $-W_1(hm, ml^-)$ for P_1 and it contributes 0 to $W_1(d, ml^-) - W_1(hm, ml^-)$ for $-\bar{P}_1$. Consequently,

$$P_1 - \bar{P}_1: -l_a(P_1).$$

There are no contributions for $-P_2$ and $\bar{P}_2$ because the foots of the crossings 12 and 14 are in ml^+ instead of ml^-.

24 contributes to $-W_1(ml, d^-)$ for P_3 and 34 contributes to $-W_1(hm, ml^+)$ for $-\bar{P}_3$ and they cancel out together.

34 contributes to $-W_1(hm, ml^+)$ for $-P_4$ and there are no contributions for $\bar{P}_4$. Consequently,

$$-P_4 + \bar{P}_4: l_a(P_4).$$

It follows from Lemma 3.1 that $l_a(P_1) = l_a(P_4)$. Consequently, $\bar{R}_a^{(2)} = 0$ on the meridian.

IV_7: $\alpha = \beta = 0$, $\gamma = n$. P_2 is of type $r(0, 0, n)$ and P_3 is of type $r(0, n, 0)$, $\infty = 4$. 13 contributes to $-W_1(ml, d^-)$ for $-P_2$. 12 contributes to $-W_1(ml, d^-)$ for $\bar{P}_2$ and they cancel out together.

There are no contributions from crossings in the tetrahedron to P_3 and $-\bar{P}_3$.

Consequently, $\bar{R}_a^{(2)} = 0$ on the meridian.

Global type V

V_1: $\alpha = a$, $\beta = 0$ and $\gamma = n - a$. Then, P_1 is of type $l(a, 0, a)$, P_3 and P_4 are of type $r(a, n, a)$ and P_2 is of type $l(n, 0, n)$, $\infty = 2$. $[23] = [24] = n$, $[34] = 0$ and hence all f-crossings change. All other crossings in the tetrahedron are a-crossings.

There are no f-crossings in P_1 and $\bar{P}_1$.

$P_3 - \bar{P}_3$: $W_1(34, 23^+) - W_1(24, 23^+)$. Note that $(24, 34)$ does not contribute because the foot of 34 is in d^-.

$P_4 - \bar{P}_4$: $W_1(23, 24^+) + 1$.

The "1" here comes from the contribution $(23, 34)$.
$l_a(P_2) = l_a(\bar{P}_2) + 1$ from the a-crossing 12. Consequently,

$P_2 - \bar{P}_2$: $W_1(24, 23^+) - W_1(34, 23^+) - (W_1(23, 24^+) + 1)$.

The "1" here comes from the fact that we have to consider the contributions on the positive side of $l(n, 0, n)$, where $(23, 34)$ contributes.

All contributions cancel out together and
$\bar{R}_a^{(2)} = 0$ on the meridian.

V_2: $\alpha = 0$, $\beta = a$ and $\gamma = n - a$. Then, P_4 is of type $r(a, n, a)$. There is no stratum $l(n, 0, n)$ and the weights cannot change. $[23] = n - a$ and hence f-crossings do not change. It follows that $R_a^{(2)} = 0$ on the meridian.

If $\gamma = n - a$, $\alpha + \beta = a$ and $\alpha \neq 0$ and $\alpha \neq a$, then P_4 is still of type $r(a, n, a)$. But there aren't any f-crossings and any other persistent crossings in the tetrahedron.

V_3: $\alpha + \gamma = a$, $\beta = 0$. Then, P_2 is of type $l(a, 0, a)$. If $\gamma = 0$, then also P_1 is of type $l(a, 0, a)$. But f-crossings do not change. If $\alpha = 0$, then P_1 is of type $l(0, 0, 0)$, which does not contribute.

It remains to consider the degenerate tetrahedrons.

V_4: $\alpha = 0$, $\beta = 0$ and $\gamma = 0$. Then, P_3 and P_4 are of type $r(0, 0, n)$, $\infty = 3$. But there aren't any contributions of crossings of the tetrahedron to any W_1.

Consequently, $\bar{R}_a^{(2)} = 0$ on the meridian.

V_5: $\alpha = n$, $\beta = 0$ and $\gamma = 0$. Then, P_1 and P_2 are of type $l(n, 0, n)$, $\infty = 2$. There aren't any contributions in P_1 and $-\bar{P}_1$ because the foots of 23 and 24 are in ml^-. 13 contributes to $W_1(hm, ml^+)$ for P_2 and 14 contributes to $W_1(hm, ml^+)$ for $-\bar{P}_2$. They cancel out together.

Consequently, $\bar{R}_a^{(2)} = 0$ on the meridian.

V_6: $\alpha = 0$, $\beta = n$ and $\gamma = 0$. Then, P_1 and P_2 are of type $l(n, n, 0)$. P_3 is of type $r(0, 0, n)$, $\infty = 1$. There aren't any contributions of crossings of the tetrahedron to any W_1 for P_1, $-\bar{P}_1$, P_2, $-\bar{P}_2$, P_3 and $-\bar{P}_3$.

Consequently, $\bar{R}_a^{(2)} = 0$ on the meridian.

V_7: $\alpha = 0$, $\beta = 0$ and $\gamma = n$. Then, P_2 is of type $l(n, 0, n)$ and P_3 and P_4 are of type $r(0, n, 0)$, $\infty = 4$. 13 contributes to $-W_1(ml, d^+)$ and 14 contributes 0 to $W_1(d, ml^+) - W_1(ml, d^+)$ for P_2. 14 contributes to $-W_1(ml, d^+)$ and 12 contributes 0 to $W_1(d, ml^+) - W_1(ml, d^+)$ for $-\bar{P}_2$. They cancel out together.

14 contributes to $-W_1(hm, ml^-)$ for P_3. 24 contributes to $-W_1(ml, d^-)$ for $-\bar{P}_3$. They cancel out together.

23 contributes 0 to $W_1(d, ml^-) - W_1(ml, d^-)$. There are no contributions for $-\bar{P}_4$.

Consequently, $\bar{R}_a^{(2)} = 0$ on the meridian.

Global type VI

VI_1: $\alpha = 0$, $\beta = a$ and $\gamma = 0$. Then, P_1 and P_2 are of type $l(a, 0, a)$. The weights cannot change and there aren't any f-crossings in the tetrahedron. Consequently, $\bar{R}_a^{(2)} = 0$ on the meridian.

VI_2: $\alpha = a$, $\beta = 0$ and $\gamma = 0$. Then, P_1, P_3 and P_4 are of type $l(a, 0, a)$ and P_2 is of type $l(0, 0, 0)$, $\infty = 3$. The weights cannot change and f-crossings do not change for P_1 and $-\bar{P}_1$.

$-P_3 + \bar{P}_3$: $-W_1(24, 23^-) + W_1(34, 23^-)$
$-P_4 + \bar{P}_4$: $W_1(23, 24^-)$.

Hence, the total contribution is

$$-W_1(24, 23^-) + W_1(34, 23^-) + W_1(23, 24^-).$$

But the three crossings form the stratum $P_2 = 234$. It follows from a figure completely analogous to Fig. 3.42 that

$-W_1(24, 23^-) + W_1(34, 23^-) + W_1(23, 24^-) = 0$ for the triple crossing $P_2 = 234$.

Consequently, $\bar{R}_a^{(2)} = 0$ on the meridian.

VI_3: $\alpha = a$, $\beta = 0$ and $\gamma = $ arbitrary, but $\gamma \neq 0$. Then, P_3 is of type $l(a, 0, a)$. But then $[24] = \beta + \gamma$ and $[34] = \gamma$ are not persistent crossings.

Consequently, $\bar{R}_a^{(2)} = 0$ on the meridian.

VI_4: $\alpha = n - a$, $\beta = a$ and $\gamma = 0$. Then, P_1 is of type $l(n, 0, n)$ and P_2 is of type $l(a, 0, a)$, $\infty = 2$ or $\infty = 3$. But P_2 contributes only for $\infty = 2$. Moreover, only for $\infty = 2$ the crossings 23 and 24 would be a-crossings. 13 and 14 are 0-crossings.

But 23 and 24 do not contribute neither to $l_a(P_1)$ nor to $l_a(\bar{P}_1)$. Consequently, $\bar{R}_a^{(2)} = 0$ on the meridian.
It remains to consider the degenerate tetrahedrons.

VI_5: $\alpha = 0$, $\beta = 0$ and $\gamma = 0$. Then, all strata are of type $l(0, 0, 0)$ and they do not contribute.

VI_6: $\alpha = n$, $\beta = 0$ and $\gamma = 0$. Then, P_1, P_3 and P_4 are of type $l(n, 0, n)$, $\infty = 3$.

24 contributes to $-W_1(ml, d^+)$ for P_1. 24 contributes 0 to $W_1(d, ml^+) - W_1(ml, d^+)$ and 23 contributes to $-W_1(ml, d^+)$. They cancel out together.

24 contributes 0 to $W_1(d, ml^+) - W_1(ml, d^+)$ for $-P_3$ and there is no contribution for $\bar{P}_3$.

13 contributes to $-W_1(hm, ml^+)$ for $-P_4$ and 23 contributes to $-W_1(ml, d^+)$ for $\bar{P}_4$. They cancel out together.

Consequently, $\bar{R}_a^{(2)} = 0$ on the meridian.

VI_7: $\alpha = 0$, $\beta = n$ and $\gamma = 0$. Then, P_1 and P_2 are of type $l(n, 0, n)$ and P_3 and P_4 are of type $l(n, n, 0)$, $\infty = 2$.

24 contributes to $-W_1(hm, ml^+)$ for P_1 and 23 contributes to $-W_1(hm, ml^+)$ for $-\bar{P}_1$. They cancel out together.

There are no contributions for P_2 and $-\bar{P}_2$ because 13 and 14 have their foots in ml^-.

34 contributes 0 to $W_1(d, ml^-) - W_1(hm, ml^-)$ for $-P_3$ and there is no contribution for $\bar{P}_3$.

There is no contribution for $-P_4$ and 34 contributes 0 to $W_1(d, ml^-) - W_1(hm, ml^-)$ for $\bar{P}_4$.

Consequently, $\bar{R}_a^{(2)} = 0$ on the meridian.

VI_8: $\alpha = 0$, $\beta = 0$ and $\gamma = n$. Then, P_1, P_2 and P_4 are of type $l(n, n, 0)$, $\infty = 1$. There aren't any contributions from crossings of the tetrahedron to the weights for any stratum.

Consequently, $\bar{R}_a^{(2)} = 0$ on the meridian.

We have proven that our 1-cochain $\bar{R}_a^{(2)}$ satisfies (b) the positive global tetrahedron equation. Sorry, but we are conscious that this proof gives a hard time to the reader.

Let us comment a bit on the *structure of this proof*, which makes it perhaps easier for the reader to understand it:

- For the global types I and VI (where all triple crossings are of type r, respectively, l), it happens that all three possible f-crossings (namely, 24, 34 and 23) change simultaneously. In this case, they form together the stratum P_2 of type $r(n, n, n)$, respectively, $l(0, 0, 0)$, and the contributions cancel out together.

- For the global types II–V, it happens too that all three possible f-crossings (namely, 24, 34 and 23) change simultaneously. Then, they would not cancel out for the degenerate triple crossing P_2. However, there is always $l_a(P_2) = l_a(\bar{P}_2) + 1$ and the contribution of the changing f-crossings cancels out with this difference between P_2 and $\bar{P}_2$.

- Exactly for the global types II and IV, it happens that in the case of a degenerate tetrahedron, a crossing contributes for P_1 but not for $\bar{P}_1$ and then a (perhaps different) crossing contributes to P_4 and not to $\bar{P}_4$. The contributions cancel out together if and only if $l_a(P_1) = l_a(P_4)$. But this is exactly the claim of Lemma 3.1.
- The set of global types of degenerate triple crossings splits into three subsets: $r(n, n, n)$ and $l(0, 0, 0)$ which do not contribute at all; $r(0, 0, n)$ and $l(n, n, 0)$ for which the individual contributions of the three crossings after a small perturbation do not depend on the side of the discriminant of the triple crossing in the moduli space; $r(0, n, 0)$ and $l(n, 0, n)$ for which the individual contributions of the three crossings after a small perturbation depend on the side of the discriminant of the triple crossing in the moduli space.

Exactly, the cases III_1 and V_1 force us to consider the contributions on the positive side of the triple crossing (i.e., exactly the couple (hm, ml) contributes, compare the co-orientation for triple points of local type 1).

After all, this proof looks to us as a bit like a miracle. We have no idea what could be its differential geometric counterpart in the infinite dimensional moduli space M_n.

$\bar{R}_a^{(2k)}$ *satisfies the cube equations (c).*

Here, we can consider again the 1-cochain $\bar{R}_a^{(2k)}$ in its whole generality and not only for $k = 1$. The proof is rather similar to the proof for $R_{a,d^+,hm}^{(2k)}$ and for $R_{a,d^+}^{(2k)}$. But we now have to consider all edges of the cube Γ because much more global types of triple crossings contribute and for the degenerate triple crossings all three crossings always contribute. Compare Fig. 2.2 for the names of local types of triple crossings.

Some of the edges are already shown at the end of Section 3.2. All edges are contained in Figs. 6.28–6.65 in Chapter 6.

Exactly as for $R_{a,d^+,hm}^{(2k)}$, the n-crossing hm in $r(a, n, a)$ always contributes with a positive sign by $W_{2k-1}(hm)w(hm)$ and the last foot of a 0-crossing is always automatically in hm^+. Figure 3.35 shows that the new negative n-crossing does not contribute for the local type 1 (i.e., all three crossings in the move are positive), but the positive n-crossing contributes for the local type 7 exactly by $W_{2k-1}(hm)w(hm)$. This is compensated in the formula for $\bar{R}_a^{(2k)}$ by the correction term $1/2(w(hm) - 1)\bar{W}_1(hm)w(hm)$ because now $w(hm) = -1$. Going through Figs. 3.35–3.38, we see that in each case with the correction term $1/2(w(hm) - 1)W_{2k-1}(hm)w(hm)$ the contributions of the two R III moves in the edge are now the same.

Figure 6.34 shows that the new negative 0-crossing contributes $-W_{2k-1}(hm)w(hm)$ to the symbol in the formula for the triple crossing of local type 1. For the local type 7, it does no longer contribute, but this is compensated in the formula by the correction term $1/2(w(hm) - 1)\bar{W}_1(hm)w(hm)$ because now $w(hm) = -1$. Going through Figs. 6.40, 6.42 and 6.46, we see that in each case with the correction term $1/2(w(hm) - 1)W_{2k-1}(hm)w(hm)$ the contributions of the two R III moves in the edge are again the same.

It remains to consider the degenerate triple crossings.

The refined linking numbers $l_a(p)$ are the same for the two R III moves in an edge because the a-crossings are not persistent crossings, and hence they cannot become crossings in the triangle.

Let us consider Fig. 3.35 with $\infty = b$. Then, the R III moves are of global type $r(0, 0, n)$. The negative crossing does not contribute to any weight for the triple crossing of local type 1 and the positive crossing does not contribute to any weight for the triple crossing of local type 7. The weight $W_{2k-1}(hm)w(hm)$ with $w(hm) = 1$ is replaced by the weight with $w(hm) = -1$ and hence it doesn't change. Evidently, the same is true for all local types of all triple crossings of the two types $r(0, 0, n)$ and $l(n, n, 0)$.

It remains to consider the global types $r(0, n, 0)$ and $l(n, 0, n)$. We will study this last type. Type $r(0, n, 0)$ is completely analogous, and we left the verifications to the reader.

Let us start by making a connection between the globally defined co-orientation of an R III move and its local type by considering the edges of the cube Γ.

For the local type 1, the positive side is characterised by the fact that the crossing d corresponds to an isolated chord in the corresponding chord diagram. It follows that

local type 2: on the positive side all three chords intersect

local type 3: on the positive side ml is the isolated chord

local type 4: on the positive side hm is the isolated chord
local type 5: on the positive side hm is the isolated chord
local type 6: on the positive side all three chords intersect
local type 7: on the positive side ml is the isolated chord
local type 8: on the positive side d is the isolated chord.

Of course, exactly as for the global types $r(0,0,n)$ and $l(n,n,0)$, we have to always consider $W_{2k-1}(c)w(c)$ for $c = d$, $c = hm$ and $c = ml$.

Let us consider the edge in Fig. 6.34. The $R\,III$ moves are of global type $l(n,0,n)$ for $\infty = a$ (as for all other edges of Γ too). For the $R\,III$ move of local type 1 on the positive side (i.e., d is isolated), we have $W_{2k-1}(d, ml^+) = W_{2k-1}(hm, ml^+) = W_{2k-1}(ml, d^+) = 0$. For the local type 7, the crossing ml is isolated on the positive side and we obtain $W_{2k-1}(d, ml^+) = W_{2k-1}(hm, ml^+) = W_{2k-1}(ml, d^+) = 0$ too.

Let us consider the edge in Fig. 6.28. On the positive side for the local type 1, we have $-W_{2k-1}(ml, d^+)w(ml) = -1$. For the local type 5, the crossing hm is isolated on the positive side and we have $-W_{2k-1}(hm, ml^+)w(ml) = -1$.

Let us consider the edge in Fig. 6.48. On the positive side for the local type 1, we have $-W_{2k-1}(hm, ml^+)w(hm) - W_{2k-1}(ml, d^+)w(ml) = 1 - 1 = 0$. On the negative side for the local type 6 (which is non-braid-like), none of the crossings intersects any other.

Let us consider the edge in Fig. 6.38. On the positive side for the local type 7, all weights are 0, as well as on the positive side for the local type 4.

Let us consider the edge in Fig. 6.30. On the positive side for the local type 7, all weights are 0. On the negative side of the local type 2 (non-braid-like), there aren't any intersections of crossings.

Let us consider the edge in Fig. 6.36. On the positive side for the local type 5, the crossing hm is isolated. There are no contributions. On the positive side for the local type 3, the crossing ml is isolated. There are no contributions neither.

Let us consider the edge in Fig. 6.46. On the positive side of the local type 5, there aren't any contributions. On the negative side of the local type 2, we have $W_{2k-1}(d, ml^+)w(d) - W_{2k-1}(ml, d^+)w(ml) = 1 - 1 = 0$.

Let us consider the edge in Fig. 6.32. On the positive side of the local type 3, there aren't any contributions. On the negative side of the local type 6 (non-braid-like), there aren't any contributions neither.

Let us consider the edge in Fig. 6.40. On the positive side of the local type 4, there aren't any contributions. On the negative side of the local type 6 (non-braid-like), we have $W_{2k-1}(d, ml^+)w(d) - W_{2k-1}(ml, d^+)w(ml) = -1 + 1 = 0$.

Let us consider the edge in Fig. 6.42. On the positive side of the local type 3, we have that $-W_{2k-1}(ml, d^+)w(ml)$ contributes $W_{2k-2}(hm, ml)w(hm)$ (where $w(hm) = +1$). But on the positive side of the local type 8, where the crossing d is isolated, $-W_{2k-1}(ml, d^+)w(ml)$ contributes also $W_{2k-2}(hm, ml)w(hm)$ (where $w(hm) = -1$ now).

Let us consider the edge in Fig. 6.51. On the positive side of the local type 8, we have $-W_{2k-1}(hm, ml^+)w(hm) - W_{2k-1}(ml, d^+)w(ml) = -1 + 1 = 0$. On the negative side for the local type 2 (which is non-braid-like), none of the crossings intersects any other.

Let us consider the edge in Fig. 6.44. On the positive side of the local type 4, there aren't any contributions. On the positive side for the local type 8, we have $-W_{2k-1}(hm, ml^+)w(hm) - W_{2k-1}(ml, d^+)w(ml) = -1 + 1 = 0$.

We have checked all 12 edges of Γ. This finishes the proof that $\bar{R}_a^{(2k)}$ satisfies the cube equations (c), and hence, it is a 1-cocycle in M_n at least for $k = 1$.

The example in Chapter 4 shows that $\bar{R}_a^{(2)}$ represents a non-trivial cohomology class in M_n. This finishes the proof of Theorem 2.5. $\square$

3.4 Proof of Theorem 2.6

We will consider only the case of $R_{a_i,d^+}^{(2k)}$. The considerations for all other combinatorial 1-cocycles are completely analogous and are left to the reader.

$R_{a_i,d^+}^{(2k)}$ satisfies evidently the commutation relations and the cube equations because the two a-crossings in $r(a, n, a)$ have the same foot and two new a-crossings from a Reidemeister II move have always the same foot too.

We have to only check that $R^{(2k)}_{a_i,d+}$ is still a solution of the tetrahedron equation.

By inspecting Fig. 3.3, we see immediately that if the tetrahedron contains a stratum of type $r(n, n, n)$, then for all strata of type $r(a, n, a)$ in the tetrahedron all the a-crossings have the same foot (namely, on branch 1 for the tetrahedron).

We go now again through the proof for $R^{(2k)}_{a,d+}$, but we have to consider only the cases where at least two strata of type $r(a, n, a)$ or an $(n - a)$-crossing in the refined linking number $l(p)$ for the type $r(n, n, n)$ occur. The contributing $(n - a)$-crossings for degenerate tetrahedrons are of course always the same for P_i and $\bar{P}_i$.

I_1: $\alpha = a$, $\beta = n - a$ and hence $\gamma = 0$. Then, P_1, P_3 and P_4 are of type $r(a, n, a)$ and P_2 is of type $r(n, n, n)$. There are no $(n - a)$-crossings, and all a-crossings have the same foot on branch 1.

I_7: $\alpha =$ arbitrary, $\beta = 0$ and $\gamma = 0$. Then, P_3 is of type $r(n, n, n)$. If $\alpha = n - a$, then $[24] = [14]$ and they contribute simultaneously to $l(p)$. But they have the same head, namely, on branch 4, and they cancel out together.

II_1: $\alpha = a$, $\beta = 0$ and $\gamma = n - a$. Then, P_1 and P_4 are of type $r(a, n, a)$. But all foots of the a-crossings are in branch 1.

II_2: $\alpha = 0$, $\beta = a$ and $\gamma = n - a$. Then, P_1 and P_2 are of type $r(a, n, a)$. $[12] = 0$, $[13] = a$ and hence $-l(P_1) + l(\bar{P}_1) = 0$ and $-l(P_2)+l(\bar{P}_2) = 0$. But then $-P_1 + \bar{P}_1$ as well as $-P_2 + \bar{P}_2$ cancel already out.

III_3: $\alpha = n - a$, $\beta = a$ and $\gamma = 0$. Then, P_1 is of type $r(n, n, n)$ and P_2 is of type $r(a, n, a)$. Here, an unexpected phenomenon arrives: $[13] = n$ and it follows that $-l(P_2) + l(\bar{P}_2) = 1$ (this is the *only* case in the global tetrahedron equation where this happens). But 34 is also the crossing hm in P_1, which is of type $r(n, n, n)$. $[12] = n - a$ and hence $-l(P_1) + l(\bar{P}_1) = 1$ too, but only for $\infty = 1$, because $\infty = 3$ has not the right position with respect to 12 (compare Definition 2.4). The foot of the a-crossings in P_2 is on branch 2, but the contributing $(n - a)$-crossing in $\bar{P}_1$ from the tetrahedron is the crossing 12, which has its head on branch 2 too.

III_4: $\alpha = a$, $\beta = n - a$ and $\gamma = 0$. Then, P_1 is of type $r(n,n,n)$. There aren't any other persistent crossings in the tetrahedron. $[12] = a$ and we have to distinguish two cases. If $a \neq n - a$, then $-l(P_1) + l(\bar{P}_1) = 0$ and the contributions of $-P_1$ and $\bar{P}_1$ cancel out together. If $a = n - a$ and $\infty = 1$, then $-l(P_1) + l(\bar{P}_1) = 1$. But exactly in this case P_2 is of type $r(n - a = a, n, n - a = a)$ with $\infty \in d^+$. The a-crossings in P_2 have their foot on branch 2. But the $(n - a)$-crossing 12 has its head on branch 2 too.

V_1: $\alpha = a$, $\beta = 0$ and $\gamma = n - a$. Then, P_3 and P_4 are of type $r(a, n, a)$. P_2 is of type $l(n, 0, n)$. $[34] = 0$, $[23] = [24] = n$ and hence both f-crossings change. $\infty = 1$ is not possible because $1 \in 34^+$, which is in contradiction with $[34] = 0$. We have only to consider $\infty = 2$. But the a-crossings in P_3 and in P_4 have all their foot on branch 1.

This finishes the proof. □

3.5 Proof of Proposition 2.1

$R_{(n,b_j),d^+}$ and $R_{(a_i,b_j),d^+}$ satisfy evidently the commutation relations and the cube equations. Moreover, moving a branch over or under a cusp (in an R I move) does not contribute neither (compare [16]).

We have to only go again through the proofs on pages 71 and 72 in Ref. [16] that $R_{(n,b)}$ and $R_{(a,b)}$ satisfy the tetrahedron equation and to check that this is still the case with the refinements.

$$R_{(n,b_j),d^+}.$$

Case IV
$-P_2$ contributes if and only if $\beta = b$, $\alpha = 0$ and $\gamma = n - b$. But then it cancels out with $\bar{P}_2$. We need $\infty = 1$ here.

P_3 contributes if and only if $\alpha = b$, $\beta = 0$ and $\gamma = 0$. Then, it cancels out with $-P_4$. We need $\infty = 2$. The head of the b-crossings in P_3 is on branch 3. The b-crossing which contributes in $-P_4$ is the crossing 23, which has its head on branch 3 too.

If $-\bar{P}_3$ would contribute, then $\alpha = b$, $\gamma = n - b$ and hence $\beta = b - n$, which is not possible in M_n.

If P_1 would contribute, then $\alpha = n$, $\gamma = 0$ and hence again $\beta = b - n$.

Case II

$\bar{P}_1$ contributes if and only if $\alpha = b$ and $\gamma = \beta = 0$. But then it cancels out with $-\bar{P}_4$. We need $\infty = 1$ here. The head of the b-crossings in P_1 is on branch 4. But the head of the b-crossings in $-\bar{P}_4$ is on branch 4 too.

If P_2 would contribute, then $\beta = n$, $\alpha = b$ and $\gamma = -b$, which is not possible in M_n. If $-\bar{P}_2$ would contribute, then $\beta = n$, $\gamma = n - b$ and $\alpha = b - n$, which is not possible in M_n.

$-P_3$ contributes if and only if $\beta = \gamma = 0$ and $\alpha = b$. In this case, it cancels out with $\bar{P}_3$. We need $\infty = 1$ here. The b-crossing in $-P_3$ is the crossing 34 with the head on branch 4, but the b-crossing in $\bar{P}_3$ is the crossing 24 with its head on branch 4 too.

This finishes the proof for $R_{(n,b_j),d^+}$. The considerations for the 1-cochain $R_{(a_i,n),d^+}$ are completely analogous and are left to the reader.

$$R_{(a_i,b_j),d^+}.$$

Case IV

$-P_2$ contributes if and only if $\alpha = 0$, $\beta = a + b - n$, and $\gamma = n - b$. Exactly in this case P_1 contributes too (with the second configuration in the formula). We need $\infty = 1$ here. In $-P_2$, the foot of the a-crossing is on branch 2 and the head of the b-crossing is on branch 4. In P_1, the head of the b-crossings are on branch 4 too. The contributing a-crossing is the crossing 23, which has its foot on branch 2.

If $\bar{P}_2$ contributes, then $\beta + \gamma = n$ and $\alpha + \beta + \gamma = a$. But this cannot happen in M_n because $\alpha \geq 0$.

P_3 contributes if and only if $\alpha = a + b - n$, $\beta = 0$ and $\gamma = n - a$. Exactly in this case $-P_4$ contributes with the third configuration too. We need $\infty = 2$ here. In P_3, the foot of the a-crossing is on branch 1 and the head of the b-crossing is on branch 3. In $-P_4$, the head of the a-crossings is on branch 1 too. The contributing b-crossing is the crossing 23, which has its head on branch 3.

If $\bar{P}_3$ contributes, then $\alpha = a + b - n$, $\gamma = 2n - a - b$ and $\beta = b - n$, which cannot happen in M_n (remember that we consider the open 2-simplex here, i.e., $b < n$).

There are no configurations at all in $-\bar{P}_1$ and in $\bar{P}_4$ which could contribute.

Case II

$\bar{P}_1$ contributes if and only if $\alpha = a + b - n$, $\beta = 0$ and $\gamma = n - a$. But exactly in this case $-\bar{P}_4$ contributes too. We need $\infty = 1$ here. But in both $\bar{P}_1$ and $-\bar{P}_4$ the foot of the a-crossing is on branch 1 and the head of the b-crossing is on branch 4.

If P_2 contributes with the second configuration, then $\beta = 0$ and hence it cancels out with the contribution of $-\bar{P}_2$. We need $\infty = 1$ here. The b-crossings have their head on branch 4. The contributing a-crossing in P_2 is the crossing 12 and the contributing a-crossing in $-\bar{P}_2$ is the crossing 13, and they have both their foot on branch 1.

If P_2 would contribute with the third configuration, then $\beta = a$, $\alpha + \gamma = 0$ and hence in M_n we have $\alpha = 0$ and $\gamma = 0$. Hence, $n - \beta - \gamma = b$ implies $a + b = n$, which is impossible because we are in the open 2-simplex. If $-\bar{P}_2$ would contribute with the third configuration, then $n = b$, which is also impossible.

$-P_3$ contributes with the third configuration if and only if $\beta = 0$, $n - \gamma = a$ and $\alpha + \gamma = b$. But it cancels out with the contribution of $\bar{P}_3$. We need $\infty = 1$ here. The a-crossings have their foot on branch 1. The contributing b-crossing in $-P_3$ is the crossing 34 and the contributing b-crossing in $\bar{P}_3$ is the crossing 24, and they have both their head on branch 4.

If $-P_3$ would contribute with the second configuration, then $\beta = b$, $\gamma = n - b$ and $\alpha = a + b - n$. But then, e.g., $\bar{P}_3$ contains an arrow with marking $\alpha + \beta + \gamma = a + b > n$, which is not possible in M_n. If $\bar{P}_3$ would contribute with the second configuration, then $\beta = b$, $\gamma = n - b$ and $\alpha = a - n$, which is impossible.

This finishes the proof for $R_{(a_i, b_j), d^+}$. $\qquad\square$

3.6   Proof of Theorem 2.7

$R_{i,j,k}^{(2)}$ satisfies evidently the commutation relations because the weight is defined by a single arrow in each configuration.

By inspecting Figs. 3.35–3.38, one easily sees that it satisfies the cube equations too. Indeed, the sign of hm does not enter, and hence $R_{i,j,k}^{(2)}$ does not change on the edge $r1 - 7$.

If $i \neq k$, then the crossings ml and d never contribute and $R_{i,j,k}^{(2)}$ satisfies the cube equations because the new crossings from the R II move have marking i.

If $i = k$, then, e.g., for the edge $r1 - 5$, there would be a contribution from the new i-crossing for the local type 1 and there wouldn't be a contribution for the local type 5. This forces us to consider also the contributions of ml and of d in this case. For the local type 1, the total contribution is then 0, no matter the side of the move. For the local type 5, it is 0 too because, e.g., on the positive side hm is an isolated arrow.

One easily checks now all the other edges in the same way.

If the isotopy is not regular, then we have to study the meridian of a cusp with a transverse branch (compare the Introduction). There is a single R III move in the meridian. For this triple crossing, one of the arcs hm or ml in the circle should be empty, i.e., it does not contain the point ∞ or the head of a k-crossing. This can appear exactly for the last configuration of $R_{i,j,k}^{(2)}$, and hence this triple crossing could contribute non-trivially. But in this case, the homological marking of ml has to be 0, and consequently $i = j$.

It follows that for $i \neq j$ the 1-cochain $R_{i,j,k}^{(2)}$ vanishes on the meridian.

We have to go now again through Figs. 3.5–3.16 and check the tetrahedron equations. We do not need the homological markings here but only the position of the point ∞. We drop all cases with no contribution at all.

Case I_3
$-P_1$ contributes with 23 and $\bar{P}_1$ contributes with 24.
$-P_2$ contributes with 14 and $\bar{P}_2$ contributes with 13.

Case I_4
P_4 contributes with 13 and $-\bar{P}_4$ contributes with 13 too.

Case II_2
P_3 contributes with 34 and $-\bar{P}_4$ contributes with 34 too.

Case II_4
$-P_1$ contributes with 12 and $\bar{P}_1$ contributes with 12 too.
$-P_2$ contributes with 12 and 14. P_2 contributes with 12 and 13.

Case III$_4$
 $-P_1$ contributes with 23 and $\bar{P}_1$ contributes with 24.
 $-P_2$ contributes with 14 and $\bar{P}_2$ contributes with 13.

Case IV$_3$
 $-P_1$ contributes with 23 and $\bar{P}_1$ contributes with 24.

Case V$_4$
 P_3 contributes with 34 and 14. $-\bar{P}_3$ contributes with 14 and $-\bar{P}_4$ contributes with 34.

Case VI$_2$
 P_3 contributes with 34 and $-\bar{P}_4$ contributes with 34 too.

It remains to note that the contributions which cancel out together have always the same foot, hence, we can fix the marking k. Moreover, the only cases where contributions cancel out from strata with different numbers is the case P_3 with P_4. But these two strata share the crossing ml, and hence, we can fix the markings i and j.

This finishes the proof of Theorem 2.7. $\qquad\qquad\qquad\square$

Chapter 4

The Examples

In this chapter, we give in many details some simple examples which we have calculated by hand.

Example 4.1 Let K be the long positive trefoil, with framing $w_1 = 1$ and $w_0 = 2$, and let $n = 2$ and $T = \sigma_1$. Consequently, $a = 1$. We consider the loop push(T) (pushing T through $2K$ in the solid torus corresponds to the loop, which is induced by the rotation of the solid torus from the cabling operation around its core).

One easily sees that there is only a single R III move p which contributes to $\nabla R_{a=1,d^+}$, $\nabla R_{a=1,ml^-}$, $\nabla R_{a=1,d^+,hm}$ and $\nabla R_{a=1,ml^-,hm}$. Indeed, the only R III move of global type $r(1,2,1)$ comes from pushing σ_1 under the two branches with crossings of marking 2, and there are no moves of type $r(2,2,2)$ because the marking of σ_1 is 1. In the second passage of σ_1, we have an R III move of type $r(1,1,2)$, which does not contribute at all.

We show the knot diagram of the R III move p together with its sign and with the corresponding Gauss diagram in Fig. 4.1 (the corresponding end points have to be identified in the obvious way). All crossings are positive. We indicate those homological markings which are different from the marking 1. After pushing σ_1 once through the cable in the solid torus, the point ∞ is now on the other branch, and we have to push σ_1 a second time through the cable in order to obtain a loop. We see that for the first pushing, the R III move is of type $r(1,2,1), \infty \in d^+$ and for the second time, it is of type $r(1,2,1), \infty \in ml^-$.

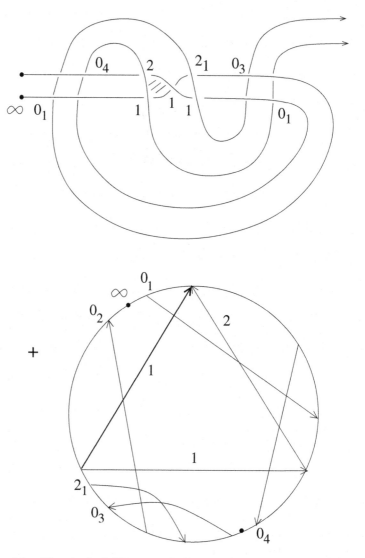

Figure 4.1:　The single R III move which contributes to $\nabla R_{a=1,d^+}(\mathrm{push}(T))$.

$W_0 = 1$ for $\nabla R_{a=1,d^+}(\mathrm{push}(T))$ and $\nabla R_{a=1,ml^-}(\mathrm{push}(T))$ because there is just one R III move which contributes and it has a positive sign.

Let us consider $\nabla R_{a=1,d^+}(\mathrm{push}(T))$ and $\nabla R_{a=1,d^+,hm}(\mathrm{push}(T))$. $W_2(p) = 1$ (contribution from 2_1 with 0_3).

$W_4(p) = 1$ (contribution from 2_1 with 0_3 together with hm with 0_1, compare the last line in Fig. 1.9. Remember that hm can contribute in the configuration but just not as the first n-crossing).

$l(p) = 0$, $W_1(hm) = 1$ (contribution from 0_1) and $W_3(hm) = 0$ (because 2_1 would be the first n-crossing and not hm).

All other W_{2k} and W_{2k-1} are 0.

It follows that $\nabla R_{a=1,d^+}(\mathrm{push}(T)) = 1 + z^2 + z^4$ and that

$$\nabla R_{a=1,d^+,hm}(\mathrm{push}(T)) = z^2.$$

Let us consider $\nabla R_{a=1,ml^-}(\mathrm{push}(T))$ and $\nabla R_{a=1,ml^-,hm}(\mathrm{push}(T))$.

$W_2(p) = 1$ (contribution from 2_1 with 0_3 because 2_1 is still the first n-crossing).

$W_4(p) = 0$ (because hm would be the first n-crossing now, which is not allowed).

$l_{ml^-}(p) = 0$, $W_1(hm) = 1$ (still contribution from 0_1) and $W_3(hm) = 1$ (because now hm is the first n-crossing in hm with 0_1 and 2_1 with 0_3).

All other W_{2k} and W_{2k-1} are 0.

It follows that $\nabla R_{a=1,ml^-}(\mathrm{push}(T)) = 1 + z^2$ and that

$$\nabla R_{a=1,ml^-,hm}(\mathrm{push}(T)) = z^2 + z^4.$$

Let us calculate the truncation of degree at most two of $\nabla|\bar{R}_{a=1}^{(2k)}(\mathrm{push}(T))$; remember that we need $\infty \in d^+$. There are no degenerate triple crossings because only the crossing σ_1 moves and it has marking 1. There are now exactly three R III moves which contribute: the same as for $\nabla R_{a=1,d^+}(\mathrm{push}(T))$, which now contributes $1 + 2z^2$ because hm is allowed to be the first n-crossing in $\bar{W}_2(p)$. One easily sees that the other two moves are of type $l(1,0,1)$ from pushing σ_1 through the crossings of marking 0. The two other R III moves together with their Gauss diagrams and their signs are shown in Figs. 4.2 and 4.3. Remember that for this type the foots of the first n-crossing and of the last 0-crossing have to be in the (closed) arc hm^-. Consequently, the move shown in Fig. 4.2 contributes $-1 - z^2$ because the foot of one of the 2-crossings is in ml^+. The move shown in Fig. 4.3 contributes $-1 - 2z^2$ because hm is allowed to be the last 0-crossing.

It follows that $\nabla|\bar{R}_{a=1}^{(2k)}(\mathrm{push}(T)) = -1 - z^2$.

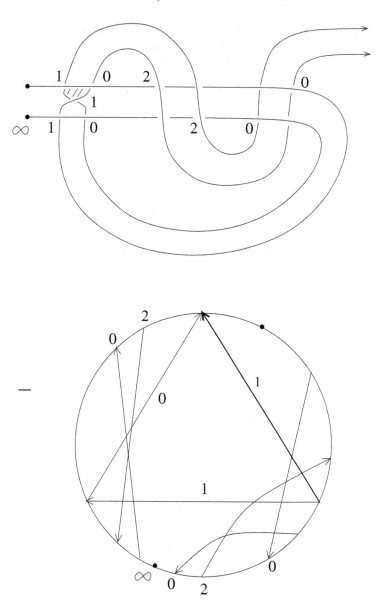

Figure 4.2: The first R III move of type $l(1,0,1)$ which contributes to $\nabla|\bar{R}_{a=1}^{(2k)}(\mathrm{push}(T))$.

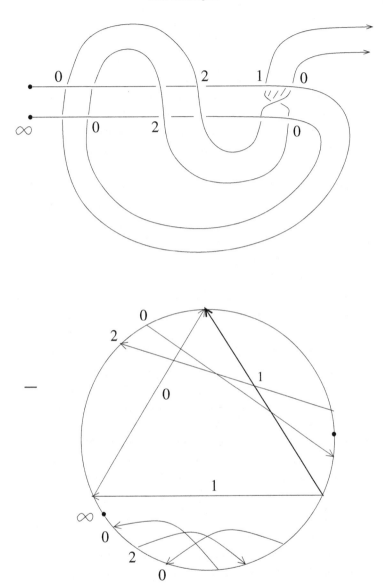

Figure 4.3: The second R III move of type $l(1,0,1)$ which contributes to $\nabla|\bar{R}_{a=1}^{(2k)}(\mathrm{push}(T))$.

This very simple example (push(σ_1) is just the rotation around the core of the solid torus from the cabling construction) already shows that all our five combinatorial 1-cocycles represent non-trivial cohomology classes in the corresponding moduli spaces.

For the positive torus knot K of type $(5,2)$, with $w_1(K) = 2$, there are exactly two moves of type $r(1,2,1)$ in the loop push(σ_1), completely analogous to the case of the positive trefoil. We have calculated that

$$\nabla R_{a,d^+}(\text{push}(\sigma_1)) = 2 + 9z^2 + 21z^4 + 12z^6 + 2z^8$$

$$\nabla R_{a,d^+,hm}(\text{push}(\sigma_1)) = 3z^2 + z^4$$

$$\nabla R_{a,ml^-}(\text{push}(\sigma_1)) = 2 + 9z^2 + 12z^4 + 6z^6 + z^8$$

$$\nabla R_{a,ml^-,hm}(\text{push}(\sigma_1)) = 3z^2 + 10z^4 + 6z^6 + z^8.$$

We left the verification to the reader. (As already mentioned, the configurations in C_{2k} for $k > 2$ which appear here in the calculations are almost all of the form of products of configurations of lower order, i.e., connected sums of configurations of lower order as in the last line of Fig. 1.9.)

But, of course, the really interesting loop is the Fox–Hatcher loop because it is only well defined for cables of long knots and it is of higher complexity than the loop from the rotation.

Example 4.2 Let K be the standard long positive trefoil with $w_1(K) = 1$ and with a positive curl with a 1-crossing added to perform the regular loop rot. Let K' be the standard long negative trefoil with $w_1(K') = 1$ and with a positive curl with a 1-crossing added to perform again the regular loop rot, compare Fig. 4.4.

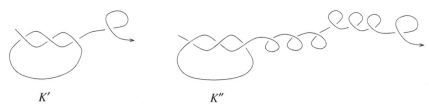

$$K' \qquad\qquad\qquad\qquad K''$$

Figure 4.4:   The two trefoils with the same writhe and the same Whitney index.

If K and K' would be isotopic knots, then the long knots K and K' would be regularly isotopic, and the oriented loops $rot(\sigma_1 \cup 2K)$ and $rot(\sigma_1 \cup 2K')$ would be homotopic in M_2. We will calculate the truncations of degree at most two $\nabla|R^{(2k)}_{a=1,d^+}(rot(\sigma_1 \cup 2K))$ and $\nabla|R^{(2k)}_{a=1,d^+}(rot(\sigma_1 \cup 2K'))$.

We consider the loop $rot(\sigma_1 \cup 2K)$.

There are exactly nine R III moves which could *a priori* contribute non-trivially to $\nabla|R^{(2k)}_{a=1,d^+}(rot(\sigma_1 \cup 2K))$. In Figs. 4.5–4.13, we show the moves, together with the Gauss diagram and the sign of the move. For an $(a\text{-})$ generic triple crossing (i.e., of global type $r(a, n, a)$, $\infty \in d^+$) as well as for the degenerate triple crossings $r(n, n, n)$, we draw only the persistent crossings in the Gauss diagram. The $[a] = [n - a] = 1$-crossings which contribute non-trivially to $l(p)$ can be established from the figures easily.

Remember that only the generic triple crossings contribute to the degree 0 term of $\nabla R^{(2k)}_{a=1,d^+}(rot(\sigma_1 \cup 2K))$. One easily sees that the result is $+3$.

Note that all generic triple crossings in this loop are of local type 1 (i.e., all three crossings are positive), and hence the correction terms from the cube equations $(w(hm) - 1)W_{2k-1}(hm)w(hm)$ are always 0.

We write then for each triple crossing the contribution to $R^{(2)}_{a=1,d^+}(rot(\sigma_1 \cup 2K))$, which can be easily checked by the reader.

$R^{(2)}_{a=1,d^+}(rot(\sigma_1 \cup 2K))$ of Fig. 4.5: contribution $+2$.

$R^{(2)}_{a=1,d^+}(rot(\sigma_1 \cup 2K))$ of Fig. 4.6: contribution $+2$.

$R^{(2)}_{a=1d^+}(rot(\sigma_1 \cup 2K))$ of Fig. 4.7: $W_1(hm) = 0$ and hence, the contribution is 0.

$R^{(2)}_{a=1,d^+}(rot(\sigma_1 \cup 2K))$ of Fig. 4.8: contribution $+1$.

$R^{(2)}_{a=1,d^+}(rot(\sigma_1 \cup 2K))$ of Fig. 4.9: we give this just as an example where a triple crossing of global type $r(a, n, a)$ does not contribute because $\infty \in ml^-$ instead of $\infty \in d^+$.

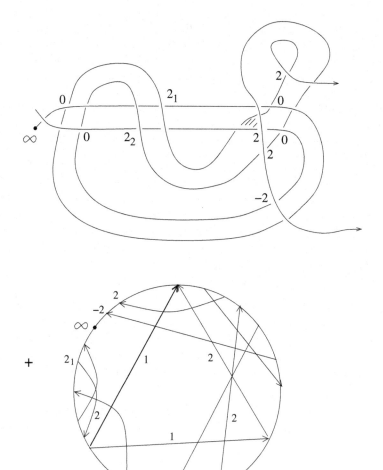

Figure 4.5: $r(1,2,1)$ move 1.

$R^{(2)}_{a=1,d^+}(rot(\sigma_1 \cup 2K))$ of Fig. 4.10: $W_1(hm) = 0$ and hence, the contribution is 0.

$R^{(2)}_{a=1,d^+}(rot(\sigma_1 \cup 2K))$ of Fig. 4.11: $W_1(hm) = 1$, but $l(p) = 0$ and hence, the contribution is 0.

$R^{(2)}_{a=1,d^+}(rot(\sigma_1 \cup 2K))$ of Fig. 4.12: contribution $-(2+1) = -3$.

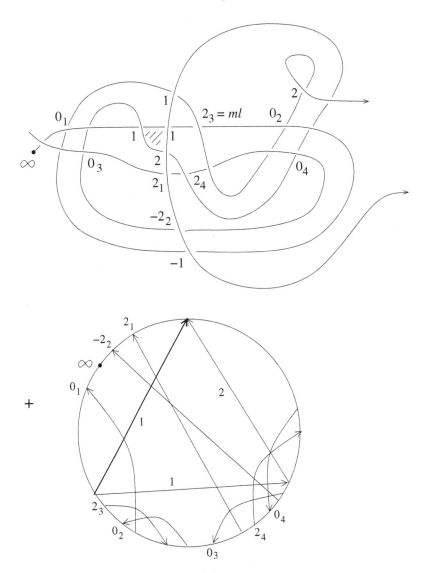

Figure 4.6: $r(1, 2, 1)$ move 2.

$R^{(2)}_{a=1,d^+}(rot(\sigma_1 \cup 2K))$ of Fig. 4.13: $W_1(hm) = 1$, and $l(p) = 2$ and hence, the contribution is $+2$ (the sign of the move is negative but this global type enters with a global minus sign into the formula).

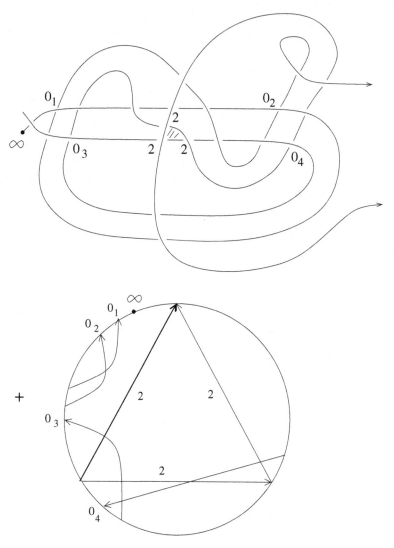

Figure 4.7: $r(2,2,2)$ move 3.

Finally, we have to push σ_1 through the 2-curl. There is only one triple crossing of the right global type in order to contribute. We show it in Fig. 4.14. Evidently, we have $l(p) = 0$. In the diagram, we see two times the knot K and it is always in hm^+. Hence, it

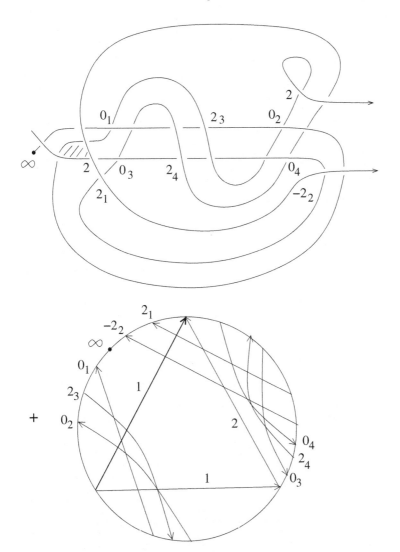

Figure 4.8: $r(1, 2, 1)$ move 4.

contributes $2v_2(K)$ to $W_2(p)$. Finally, we can push the 2-curl along the remaining part of the solid torus into its initial position without any Reidemeister moves and we obtain a loop.

$R^{(2)}_{a=1,d^+}(rot(\sigma_1 \cup 2K))$ of Fig. 4.14: contribution $+2v_2(K) = +2$.

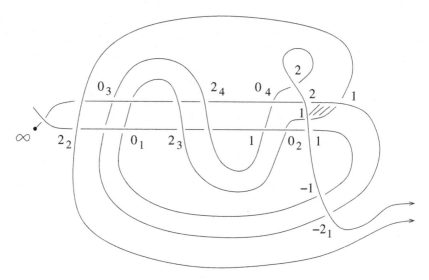

Figure 4.9: $r(1,2,1)$ with $\infty \in ml^-$.

Summing up all the contributions (we left the calculation of the higher degrees to the reader), we obtain finally

$$\nabla R_{a=1,d^+}(rot(\sigma_1 \cup 2K)) = 3 + 7z^2 + 3z^4.$$

Going again through the figures, we calculate immediately that

$$\nabla R_{a=1,d^+,hm}(rot(\sigma_1 \cup 2K)) = -z^2.$$

We consider now the loop $rot(\sigma_1 \cup 2K')$.

For the loop $rot(\sigma_1 \cup 2K')$, the calculations are longer because now there are negative crossings and the correction terms $(w(hm) - 1)$ $W_1(hm)w(hm)$ will contribute non-trivially too. Moreover, there are now exactly eight moves of the global type $r(n,n,n)$. Most of the moves are shown in Figs. 4.15–4.20. In some of the figures, we consider for simplicity simultaneously two moves. There are only two 0-crossings in the diagrams (besides in the additional 2-curls), and we skip those three of the moves of the type $r(2,2,2)$ which can evidently not contribute. Moreover, we have also to push the additional 2-curls in $2K'$ through the 2-curl, which we use to perform the loop *rot*. The 2-curls with 0-crossings can be transformed into three full-twists σ_1^2 in $M_n^{\text{semi}-\text{reg}}$. One easily sees that they contribute

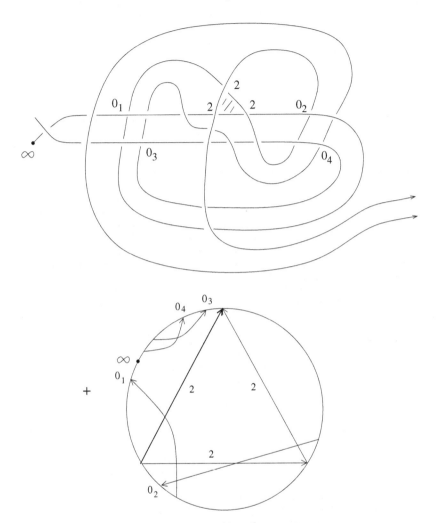

Figure 4.10: $r(2,2,2)$ move 5.

$3(1 + 2z^2)$. Pushing a 2-curl with 2-crossings through another 2-curl with 2-crossings leads just to a small loop, namely, the exchange of two 2-curls. One easily calculates that $\nabla|R^{(2k)}_{a=1,d^+}$ is 0 on this loop.

$R^{(2)}_{a=1,d^+}(rot(\sigma_1 \cup 2K'))$ of Fig. 4.15: contribution -2.

$R^{(2)}_{a=1,d^+}(rot(\sigma_1 \cup 2K'))$ of Fig. 4.16: contribution -2.

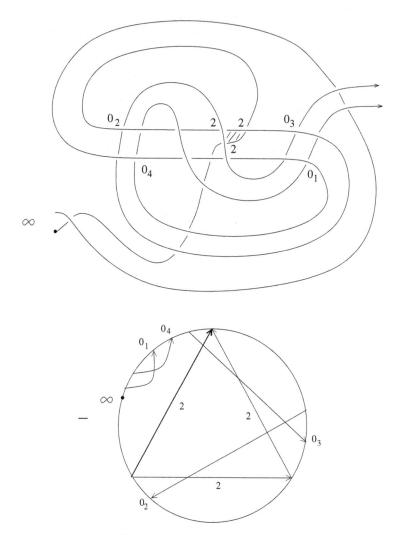

Figure 4.11: $r(2,2,2)$ move 6.

$R^{(2)}_{a=1,d^+}(rot(\sigma_1 \cup 2K'))$ of Fig. 4.17: contribution -2.

$R^{(2)}_{a=1,d^+}(rot(\sigma_1 \cup 2K'))$ of Fig. 4.18: contribution $+1$.

$R^{(2)}_{a=1,d^+}(rot(\sigma_1 \cup 2K'))$ of Fig. 4.19: contribution $+(1 + (1 - 2)$
$(-1)) = +2$.

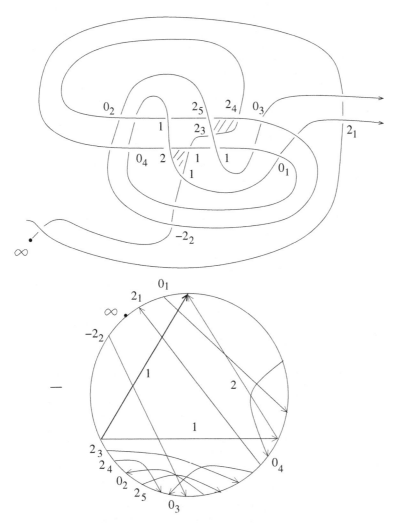

Figure 4.12: $r(1, 2, 1)$ move 7.

$R^{(2)}_{a=1,d^+}(rot(\sigma_1 \cup 2K'))$ of Fig. 4.20: contribution $W_1(hm)w$ $(hm) = -1$, $l(p) = +1$. The move has positive sign but enters with a global minus sign. Hence, the contribution is $+1$. The two moves which are shown in Fig. 4.21 cancel out together.

Pushing σ_1 through the 2-curl contributes again $1 + 2z^2$, exactly as in the case of the positive trefoil.

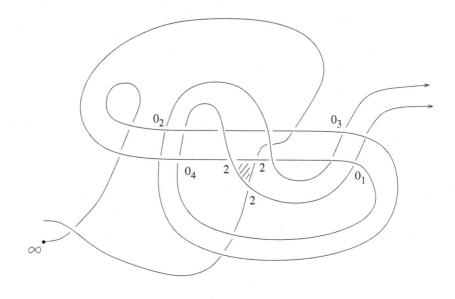

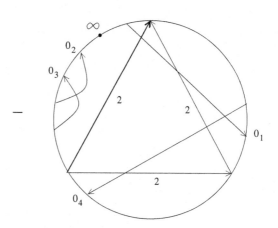

Figure 4.13: $r(2, 2, 2)$ move 8.

Summing up all the contributions, we obtain finally

$$\nabla R_{a=1,d^+}(rot(\sigma_1 \cup 2K')) = 3 + 7z^2 + 3z^4.$$

Going again through the figures, we calculate immediately that

$$\nabla R_{a=1,d^+,hm}(rot(\sigma_1 \cup 2K')) = -z^2.$$

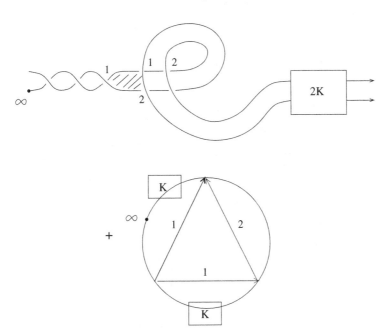

Figure 4.14: Pushing σ_1 through the 2-curl.

Consequently, $\nabla R_{a=1,d^+}(rot(\sigma_1 \cup 2K))$ does not distinguish the two trefoils, even if the 1-cocycles are very different. It could well be that the following holds: Let K be a knot with framing (writhe) $w(K)$. Then,

$$\nabla R_{a=1,d^+}(rot(\sigma_1 \cup 2K)) + \nabla R_{a=1,d^+,hm}(rot(\sigma_1 \cup 2K)) = w(K)\nabla^2_K(z).$$

This is a very particular case of the *pairing of string links with knots*:

$push(T \cup nK, nK')$. Here, T is an n-component coherently oriented string link, which induces a cyclic permutation of its end points, and nK and nK' are parallel n-cables of long framed knots K and K'. In the example $n = 2$, $T = \sigma_1$, K is the standard diagram of the positive trefoil and K' is the unknot represented by a curl with positive Whitney index and a positive crossing.

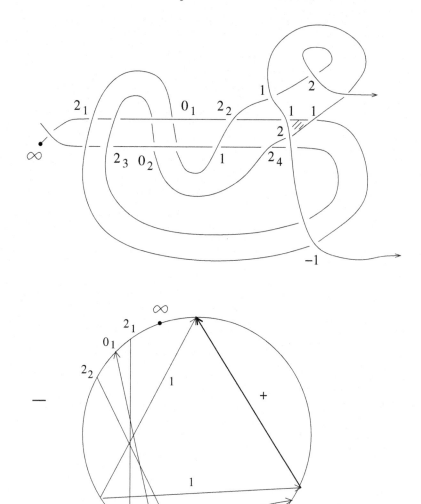

Figure 4.15: $r(1,2,1)$ and $r(2,2,2)$ moves 1 and 2.

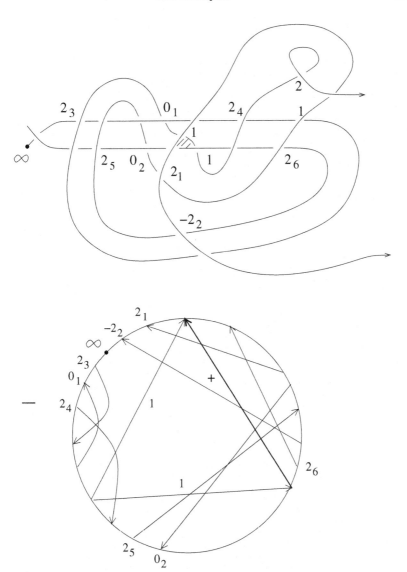

Figure 4.16: $r(1, 2, 1)$ move 3.

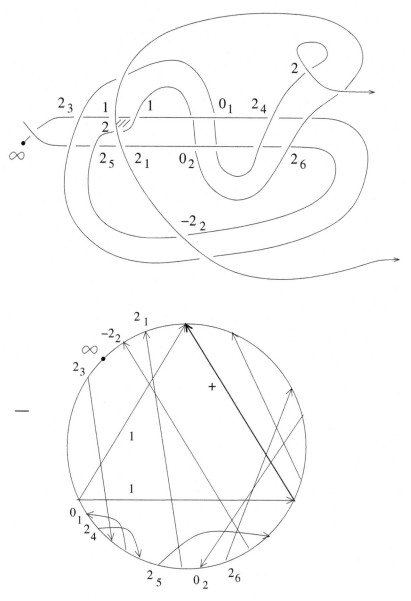

Figure 4.17: $r(1,2,1)$ and $(2,2,2)$ moves 4 and 5.

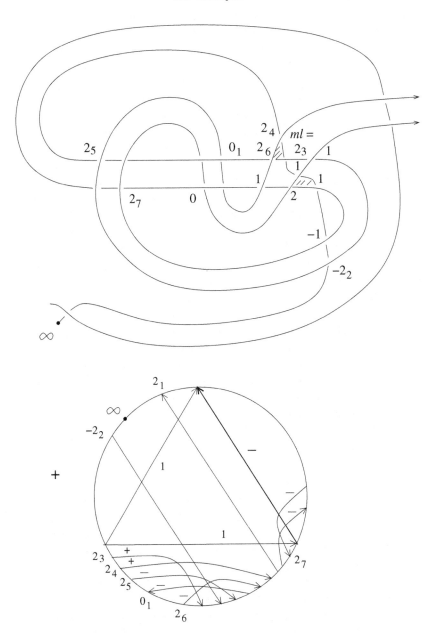

Figure 4.18: $r(1, 2, 1)$ and $r(2, 2, 2)$ moves 6 and 7.

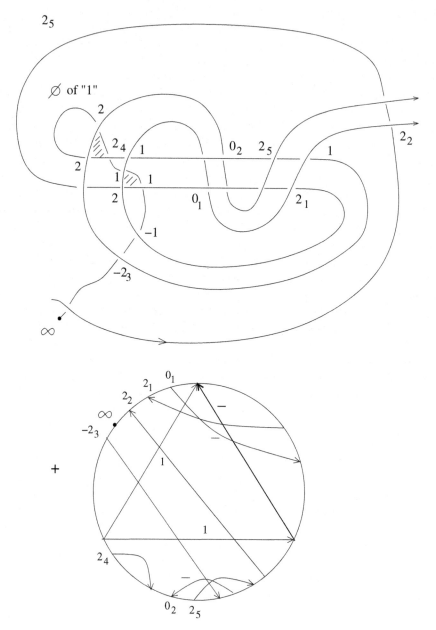

Figure 4.19: $r(1, 2, 1)$ and $(2, 2, 2)$ moves 8 and 9.

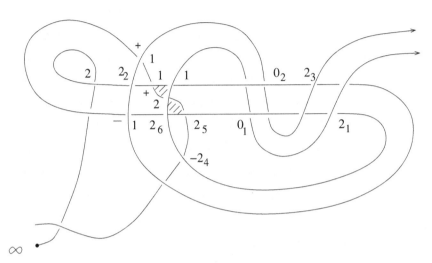

Figure 4.20: $r(2,2,2)$ move 10.

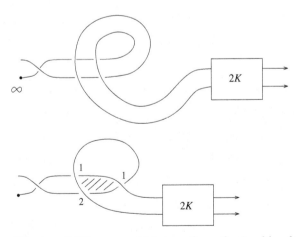

Figure 4.21: The two R III moves which transform the 2-cable of the curl into the full twist σ_1^2 do not contribute.

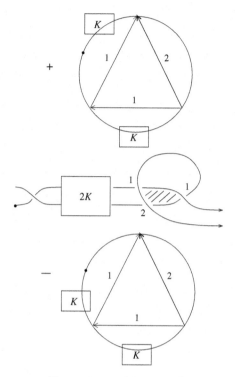

Figure 4.21: (*Continued*).

Already, if we change a single crossing in nK, then the resulting
string link $T \cup (n\tilde{K})$ would no longer be a twisted n-cable of a knot.
It is evident that the invariants of K', namely, $\nabla R_{a,d^+}(\text{push}(T \cup$
$(n\tilde{K}), nK'))$ and $\nabla R_{a,d^+,hm}(\text{push}(T\cup(n\tilde{K}), nK'))$, depend now non-
trivially on the parameter a even for $n = 2$.

It would be of course also very interesting to calculate more
1-cocycles, especially by using the splittings:

$$\nabla R_{a_i=2,d^+}(rot(\sigma_1\sigma_2 \cup 3K)), \ \nabla R_{a_i=2,d^+,hm}(rot(\sigma_1\sigma_2 \cup 3K)) \text{ and }$$

$$\nabla|\bar{R}_{a_i=2}^{(2k)}(rot(\sigma_1\sigma_2 \cup 3K)) \text{ for the two trefoils for } i = 1 \text{ and for}$$

$i = 2$ (compare Section 2.3).

But also the more complicated $\nabla R_{a=1,d^+,hm}(fh(\sigma_1 \cup 2K))$ and

$\nabla R_{a=1,d^+}(fh(\sigma_1 \cup 2K))$ and especially $\nabla|\bar{R}_{a=1}^{(2k)}(fh(\sigma_1 \cup 2K))$ for the
two trefoils would be already very interesting.

Chapter 5

Two Forgotten Linear 1-Cocycles

We define two combinatorial linear 1-cocycles, which we have missed in Ref. [16].

Let K be an arbitrary oriented knot in the solid torus with $[K] = n$, and let M be the moduli space of all knots in the solid torus which represent the given homology class n. Let M^{reg} be the corresponding moduli space of knots up to regular isotopy with respect to the projection into the annulus. We define a 1-cocycle which depends on three integer parameters: $a, b, c \in \mathbb{Z}$. Let $[a, b, c]$ denote the equivalence class of the triple (a, b, c) under cyclic permutations.

Definition 5.1 *The 1-cocycle $R_{[a,b,c]}$ is defined with the usual notation conventions (compare also Ref. [16]) by the formula given in Fig. 5.1.*

Remember that this means that we sum up over all signed triple crossings the weights (i.e., the algebraic number of crossings which correspond to the weight in the Gauss diagram) and we do not care about the signs of the three ordinary crossings in the triple crossing. The marking, e.g., $a \in H_1(V^3)$ of an arrow r means that for the corresponding crossing $[D_r^+] = a$, compare the Introduction (Chapter 1).

Proposition 5.1 *$R_{[a,b,c]}$ is a 1-cocycle in M^{reg} for all integers a, b, c. If a, b and c are neither 0 nor n, then $R_{[a,b,c]}$ is a 1-cocycle in M.*

Note that *a priori*, the 1-cocycles $R_{[a,b,c]}$ and $R_{[a,c,b]}$ are different.

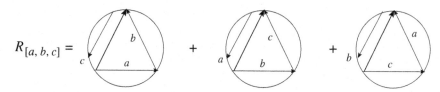

Figure 5.1:   The 1-cocycle $R_{[a,b,c]}$.

Proof.

(a) The commutation relations are evidently satisfied for each linear combinatorial 1-cochain if a, b and c are all different from 0 and n. An R I move creates a crossing of marking 0 or n. The arrow of the weights does not intersect the triangle and hence the 1-cochain does not satisfy the commutation relations with R I moves if one of a, b or c is equal to 0 or n.

(b) For the positive global tetrahedron equations, we use Figs. 3.17–3.28, which give the homological markings of the arrows. We consider the homological markings up to cyclic permutations. We give the homological markings always in the following order: *weight, ml, hm*.

Global type I: $-P_2$ contributes if and only if $n - \beta = c$, $n - \gamma = a$ and $\alpha + \beta + \gamma = b$. But then P_3 contributes too because $\alpha + \beta + \gamma = b$, $n - \beta = c$ and $n - \gamma = a$.

For all other global types, there aren't any contributions.

(c) The cube equations are evidently satisfied because the crossing of the weight can never be in an R II move together with a crossing from the triangle. Indeed, if the arrow of the weight does not cut the triangle, then it could be in an R II move only if it is oriented parallel to the corresponding crossing in the triangle.

Evidently, if neither a nor b nor c is a persistent crossing, $R_{[a,b,c]}$ satisfies the moving cusps equations. On the other hand, if one of them is a persistent crossing, then $R_{[a,b,c]}$ does not satisfy the commutation relations with R I moves.   □

Let this time (a, b, c) be an ordered triple of integers.

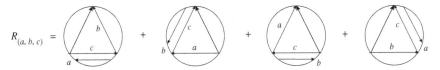

$$R_{(a,\,b,\,c)} =$$

Figure 5.2: The 1-cocycle $R_{(a,b,c)}$.

Definition 5.2 *The 1-cocycle $R_{(a,b,c)}$ is defined by the formula given in Fig. 5.2.*

Proposition 5.2 $R_{(a,b,c)}$ *is a 1-cocycle in M^{reg} for all integers a, b, c. If a, b and c are neither 0 nor n, then $R_{(a,b,c)}$ is a 1-cocycle in M.*

Proof. (a) and (c) and the discussion of $R\,\mathrm{I}$ moves are exactly the same as in the proof of the previous proposition.

(b): For the positive global tetrahedron equations, we use again Figs. 3.17–3.28, which give the homological markings of the arrows. The homological markings are now ordered because the four configurations in $R_{(a,b,c)}$ are two by two distinct even if we forget about the markings.

Global type I: No contributions at all.

Global type II: $-P_1$ contributes if and only if $\beta = a$, $n - \gamma = b$ and $n + \alpha - \beta = c$. P_2 contributes if and only if $n - \gamma = b$, $\beta = a$ and $\alpha + \gamma = c$. Consequently, they cancel out together.
    P_4 contributes if and only if $\beta = a$, $n - \beta - \gamma = c$ and $n + \alpha - c = b$. $-P_3$ contributes if and only if $\alpha + \beta + \gamma = b$, $n - \beta - \gamma = c$ and $\beta = a$. Consequently, they cancel out together.

Global type III: $\bar{P}_2$ contributes if and only if $\alpha = a$, $n - \alpha - \gamma = c$ and $\alpha + \beta + \gamma = b$. But $-P_3$ contributes if and only if $\alpha + \beta + \gamma = b$, $\alpha = a$ and $n - \alpha - \gamma = c$.

Global type IV: $\bar{P}_2$ contributes if and only if $\alpha = a$, $\alpha + \beta + \gamma = b$ and $c = n + \beta - b$. $-\bar{P}_1$ contributes if and only if $\alpha = a$, $\alpha + \beta + \gamma = b$ and $n - \alpha - \gamma = c$. They cancel out together. $-\bar{P}_3$ contributes if and only if $\beta = a$, $n - \beta - \gamma = c$ and $b = n + \alpha - c$. But $\bar{P}_4$ contributes if and only if $\alpha + \beta + \gamma = b$, $n - \beta - \gamma = c$ and $\beta = a$. They cancel out.

Global type V: $-\bar{P}_3$ contributes if and only if $\beta = a$, $\alpha + \gamma = c$ and $n - \gamma = b$. But P_2 contributes if and only if $n - \gamma = b$, $\alpha + \gamma = c$ and $\beta = a$. They cancel out.

Global type VI: No contributions at all.   □

Remark 5.1 $R_{[a,b,c]}$ *and* $R_{(a,b,c)}$ *are linear 1-cocycles, i.e., the weights are defined by using just individual crossings outside the triangle corresponding to the R III move. They depend on three homological parameters.*

In Ref. [16], we have constructed lots of linear 1-cocycles, which depend on a single homological parameter. They all turned out to represent non-trivial cohomology classes. But there was also defined a 1-cocycle, called $R_{(a,b)}$, which depends on two homological parameters. Roland van der Veen has calculated $R_{(a,b)}(rot(T \cup nK)$ for many classical knots K with his program, and the result was always 0.

I have calculated $R_{[a,b,c]}$ and $R_{(a,b,c)}$, for a, b and c neither 0 nor n, by hand for some simple examples and the result was always 0 too. It seems likely that they represent both the trivial cohomology classes. Hence, we can consider these 1-cocycles as combinatorial identities. Especially $R_{[a,b,c]} = 0$ looks to us as some kind of Jacobi identity.

Of course, taking mirror images of everything leads to other 1-cocycles.

Chapter 6

An Eclectic 1-Cocycle

This chapter contains some previous work. It was a necessary step in order to understand better how to construct combinatorial 1-cocycles. Our work has finally culminated in the combinatorial 1-cocycles, which are contained in Chapter 2 of this book. We add this chapter just for completeness.

We give now a method to construct non-symmetric solutions of the tetrahedron equation from solutions of the Yang–Baxter equation. The solution in the HOMFLYPT case gives rise to a "combinatorial quantum 1-cocycle" which represents a non-trivial cohomology class in the topological moduli space of long knots, see Section 6.11. However, its construction does not use the homological markings (as in the previous chapters) and it uses the point at infinity only in a very weak way.

Surprisingly, the formula for the solution in the HOMFLYPT case of the positive tetrahedron equation also gives a solution in the case of the 2-variable Kauffman invariant. But there is also a second non-expected solution, which gives rise to yet another non-trivial quantum 1-cocycle.

Let K be a long knot and let M_K be its topological moduli space. In this chapter, we construct a combinatorial 1-cocycle for M_K which is based on the HOMFLYPT invariant, see Theorem 6.3 in Section 6.11. It is called $\bar{R}^{(1)}$ ("R" stands for Reidemeister).

The HOMFLYPT polynomial is defined by the skein relations shown in Fig. 6.1. It is normalised as usual by $v^{-w(D)}$, where $w(D)$ is the writhe of the diagram D of the knot K.

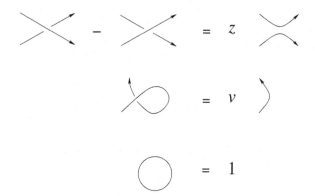

Figure 6.1:   Skein relations for the HOMFLY polynomial.

Conjecture 6.1 *Let K be a long knot, let $v_2(K)$ be its Vassiliev invariant of degree 2 and let P_K be its HOMFLYPT polynomial. Let δ denote as usual the HOMFLYPT polynomial of the trivial 2-component link. Then,*

$$\bar{R}^{(1)}(rot(K)) = -\delta v_2(K)P_K.$$

We have proven this conjecture for the trivial knot, the trefoil and the figure eight knot (Section 6.11, Example 6.3 but compare also Remark 6.11 in Section 6.11).

On the other hand, let K be the figure eight knot. A calculation by hand gives $\bar{R}^{(1)}(fh(K)) = 0$. Consequently, we have re-proven that the two loops rot and fh for the figure eight knot are not homologous (and hence in particular the figure eight knot is hyperbolic) by using only quantum topology instead of deep results in 3-dimensional topology.

We construct also another 1-cocycle, called $R_{\text{reg}}^{(1)}$, which is well defined only for regular isotopies (i.e., Reidemeister moves of type I are not allowed). The corresponding topological moduli space is called M_K^{reg}.

In fact, our 1-cocycle $\bar{R}^{(1)}$ is the result of gluing together two 1-cocycles which are of a completely different nature. First, we construct a quantum 1-cocycle $R^{(1)}$ (i.e., its construction uses skein relations of quantum invariants) which is well defined only in the complement of certain strata of codimension two in M_K which correspond to diagrams which have in the projection to the plane a branch which passes transversally through a cusp. Luckily, the value of $R^{(1)}$ on the meridians of these strata is just an integer multiple of δP_K. Next, we construct an integer-valued finite type Gauss diagram 1-cocycle $V^{(1)}$ (i.e., it is constructed by Gauss diagram formulas of finite degrees) in the complement of exactly the same strata of codimension two. We define then the completion $\bar{R}^{(1)} = R^{(1)} - \delta P_K V^{(1)}$ which turns out to be a 1-cocycle in the whole topological moduli space M_K.

All quantum invariants verify *skein relations* and hence give rise to *skein modules*, i.e., formal linear combinations over some ring of coefficients of tangles T (with the same oriented boundary ∂T) modulo the skein relations. The multiplication of tangles is defined by their composition (if possible). The main property of quantum invariants is their *multiplicativity*. We denote the HOMFLYPT skein module associated with the oriented boundary of a tangle T by $S(\partial T)$.

We will consider only a special class of tangles. *n-string link T* are n ordered oriented properly and smoothly embedded arcs in the 3-ball. They are considered up to ambient isotopy which is the identity on the boundary of the 3-ball. They generalise long knots. We chose an abstract closure of the string link to a circle.

So, a tangle is for us a knot with only a part of it embedded in 3-space.

It is well known that if the complement of the string link T does not contain an incompressible torus, then the topological moduli space M_T is a contractible space. This is always the case for braids.

Hence, there aren't any non-trivial cohomology classes in this case. However, our 1-cocycles $\bar{R}^{(1)}$ and $R_{\text{reg}}^{(1)}$ have a remarkable property, which we call the *scan-property*. We fix an orthogonal projection of the 3-ball into a disc such that the string link T is represented by a generic diagram. We fix an arbitrary abstract closure σ of T to an oriented circle, and we fix a *point at infinity* in ∂T. Let $R^{(1)}$ be a combinatorial 1-cocycle in M_T^{reg} or in M_T (i.e., we sum up contributions from Reidemeister moves).

Definition 6.1 *The 1-cocycle $R^{(1)}$ has the* scan-property *if the contribution of each Reidemeister move t doesn't change when a branch of T has moved under the Reidemeister move t to the other side of it.*

Let us add a small positive curl to an arbitrary component of T near to the boundary ∂T of T.

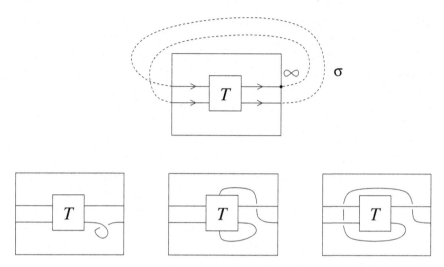

Figure 6.2: Scan-arc for a tangle T.

Definition 6.2 *The* scan-arc scan(T) *in* M_T^{reg} *is the regular isotopy which makes the small curl big under the rest of T up to being near to the whole boundary of the disc, compare Fig. 6.2 (it is convenient to replace the ball by the cube).*

Each point in ∂T comes with a sign as a component of the boundary of the oriented tangle T. We consider all cyclically ordered (by $\sigma \cup T$) sub-sets A in ∂T with alternating signs and which contain the point at infinity. The 1-cocycle $R_{\text{reg}}^{(1)}$ will be *graded* by the sets A, in contrast to the 1-cocycle $\bar{R}^{(1)}$ which cannot be graded.

Theorem 6.1 $\bar{R}^{(1)}(scan(T))$ *and* $R_{\text{reg}}^{(1)}(A)(scan(T))$ *with values in the HOMFLYPT skein module of ∂T do not depend on the chosen regular diagram of T (but they depend on the chosen closure, on the chosen component for the curl, of the choice of ∞ in ∂T and of the grading A in the case of $R_{\text{reg}}^{(1)}(A)(scan(T)))$ and they are consequently invariants of regular isotopy for the string link T.*

This means that the evaluation of the 1-cocycles on some canonical arc (instead of a loop) are already invariants. This is a consequence of the scan-property. In other words, the 1-cocycles $\bar{R}^{(1)}$ and $R_{\text{reg}}^{(1)}$ can be useful even when they represent the trivial cohomology class. Note that the scan-property is *a priori* a property of the cocycle and not of the corresponding cohomology class. (There are "dual" 1-cocycles which have the scan-property for small curls which become big over everything instead of under everything. It could well be that they represent the same cohomology class as $\bar{R}^{(1)}$, respectively, $R_{reg}^{(1)}$.)

If the long oriented knot is framed, i.e., a trivialisation (standard at infinity) of its normal bundle is chosen, then we can replace the knot K by n parallel oriented copies with respect to this framing. The result is an n-string link which is called the non-twisted n-cable $Cab_n(K)$ if the framing was zero. The loop $fh(Cab_n(K))$ is still well defined. Indeed, we put a pearl B on the braid closure of $Cab_n(K)$ in S^3 in such a way that in its interior we have n almost straight parallel lines and we push it once along the closed $Cab_n(K)$. The result is a loop in $M_{Cab_n(K)}$. We now fix an abstract closure of $Cab_n(K)$ to a circle. Consequently, $\bar{R}^{(1)}(fh(Cab_n(K)))$ is well defined.

The most important properties of our 1-cocycle $\bar{R}^{(1)}$ are the facts that it represents a non-trivial cohomology class and that it has the scan-property. Hence, it gives a new sort of calculable invariants for tangles.

Our 1-cocycle $\bar{R}^{(1)}$ allows already sometimes to answer (in the negative) a difficult question: Are two given complicated isotopies of long knots homologous in M_K? We don't even need to find out first whether these are isotopies of the same knot. For example, $rot(K)$ and $fh(K)$ are not homologous for the figure eight knot. Let us take the product of two long standard diagrams of the right trefoil. We consider the dragging of one through the other, denoted by $drag3_1^+$, see Fig. 6.3. A calculation by hand gives $\bar{R}^{(1)}(drag3_1^+) = -3\delta(P_{3_1^+})^2$ (Section 6.11, Example 6.4).

Let D be the standard diagram of the knot 3_1^+. Our calculation of $\bar{R}^{(1)}(rot(3_1^+))$ implies easily that $\bar{R}^{(1)}(rot(3_1^+\sharp3_1^+)) = -2\delta(P_{3_1^+})^2$. Consequently, we have re-proven that $dragD$ and $rot(3_1^+\sharp3_1^+)$ are not homologous for the connected sum of two right trefoils (compare Refs. [7,25]). Note that we could have altered these loops by lots of unnecessary Reidemeister moves. *There isn't known any algorithm to detect the homology class of a loop in M_K.*

More generally, let $drag(T, T')$ be the loop which consists of dragging the tangle T through the tangle T' (if possible) followed by dragging T' through T. For example, $drag(Cab_n(K), Cab_n(K'))$ is

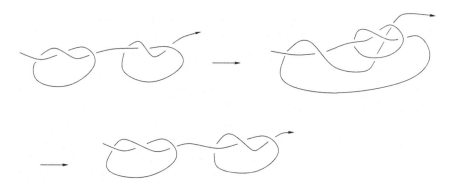

Figure 6.3:   Dragging a trefoil through another trefoil.

well defined for each couple of long knots K and K'. This is still a regular loop and we can calculate $\bar{R}^{(1)}$, $R^{(1)}_{\text{reg}}$ and $V^{(1)}$ for it. These are invariants of the couple (K, K') for each abstract closure of $Cab_n(K) \sharp Cab_n(K')$ to a circle and for each choice of a point at infinity in $\partial Cab_n(K) \sharp Cab_n(K')$. The cocycles $\bar{R}^{(1)}$ and $R^{(1)}_{\text{reg}}$ take their values in the skein module $S(\partial Cab_n(K) \sharp Cab_n(K'))$ and $V^{(1)}$ is just an integer. Are these already new knot invariants?

Remark 6.1 *In all our examples, the value of $\bar{R}^{(1)}$ on a loop contains the HOMFLYPT invariant of the knot as a factor. We don't know whether this happens in general. Let, e.g., K be a satellite and let V be the non-trivially embedded solid torus in the 3-sphere which contains K. The knot K is not isotopic to the core C of the solid torus V. We can assume that K is tangential to C at some point and we take this point as the point at infinity in the 3-sphere. The rotation of V around its core C induces now a loop for the long knot K. We call this loop $rot_C(K)$. It would be very interesting to calculate $\bar{R}^{(1)}(rot_C(K))$ in examples.*

However, the value of $\bar{R}^{(1)}$ on scan-arcs instead of loops does in general not contain the HOMFLYPT invariant as a factor (compare Examples 6.1 and 6.2 in Section 6.8)!

Question 6.1 *In particular, does $\bar{R}^{(1)}(hat(Cab_2(K)))$ contain $P_{Cab_2(K)}$ in the Hecke algebra $H_2(z, v)$ as a factor for each choice of a point at infinity in $\partial Cab_2(K)$?*

6.1 Method

For the convenience of the reader, we recall here parts of our method, which was used in the previous chapters. The construction of $\bar{R}^{(1)}$ is based on a new solution of some tetrahedron equation. The classical tetrahedron equation is a 3-dimensional generalisation of the Yang–Baxter equation. It is a local equation which is fundamental for studying integrable models in $2 + 1$-dimensional mathematical physics. This equation has many solutions and the first one was found by Zamolodchikov (see Refs. [30,33]). We look at the tetrahedron equation from the point of view of singularities of projections of lines. We fix an orthogonal projection $pr : I^3 \to I^2$. Consider four oriented straight lines in the cube I^3 which form a braid and for which the intersection of their projection into the square I^2 consists of a single point. We call this an *ordinary quadruple crossing*. After a generic perturbation of the four lines, we will now see exactly six ordinary crossings. We assume that all six crossings are positive and we call the corresponding quadruple crossing a positive quadruple crossing. Quadruple crossings form smooth strata of codimension 2 in the topological moduli space of lines in 3-space which is equipped with a fixed projection pr. Each generic point in such a stratum is adjacent to exactly eight smooth strata of codimension 1. Each of them corresponds to configurations of lines which have exactly one ordinary triple crossing besides the remaining ordinary crossings. We number the lines from 1 to 4 from the lowest to the highest (with respect to the projection pr). The eight strata of triple crossings glue pairwise together to form four smooth strata which intersect pairwise transversally in the stratum of the quadruple crossing (see, e.g., Ref. [19]). The strata of triple crossings are determined by the names of the three lines which give the triple crossing. For shorter writing, we give them names from P_1 to P_4 and $\bar{P}_1$ to $\bar{P}_4$ for the corresponding stratum on the other side of the quadruple crossing. We show the intersection of a normal 2-disc of the stratum of codimension 2 of a positive quadruple crossing with the strata of codimension 1 in Fig. 3.2. We give in the figure also the coorientation of the strata of codimension 1, which will be defined in Definition 6.1. (We could interpret the six ordinary crossings as the edges of a tetrahedron and the four triple crossings likewise as the vertices's or the 2-faces of the tetrahedron.)

Note that this coorientation is different from the one which we have used in the previous chapters: it is defined locally, but it uses the three crossings and not only the underlying planar curve.

Let us consider a small circle in the normal 2-disc which goes once around the stratum of the quadruple crossing. We call it a *meridian m* of the quadruple crossing. Going along the meridian m, we see ordinary diagrams of 4-braids and exactly eight diagrams with an ordinary triple crossing. We have shown this in Fig. 3.1. (For simplicity, we have drawn the triple crossings as triple points, but the branches do never intersect.) For the classical tetrahedron equation, one associates to each stratum P_i some operator (or some R-matrix) which depends only on the names of the three lines and to each stratum $\bar{P}_i$ the inverse operator. The tetrahedron equation says now that if we go along the meridian m, then the product of these operators is equal to the identity. Note, that in the literature (see, e.g., Ref. [30]) one considers planar configurations of lines. But this is of course equivalent to our situation because all crossings are positive and hence the lift of the lines into 3-space is determined by the planar picture. Moreover, each move of the lines in the plane which preserves the transversality lifts to an isotopy of the lines in 3-space.

However, the solutions of the classical tetrahedron equation are not well adapted in order to construct 1-cocycles for moduli spaces of knots. There is no natural way to give names to the three branches of a triple crossing in an arbitrary knot isotopy besides in the case of braids. But it is not hard to see that in the case of braids, Markov moves would make big trouble (see, e.g., Ref. [1] for the definition of Markov moves). As is well known, a Markov move leads only to a normalisation factor in the construction of 0-cocycles (see, e.g., Ref. [44]). However, the place in the diagram and the moment in the isotopy of a Markov move become important in the construction of 1-cocycles (as already indicated by the non-triviality of Markov's theorem). To overcome this difficulty, we search for a solution of the tetrahedron equation which does not use names of the branches.

The idea is to associate with a triple crossing p of a diagram of a string link T, a global element in the skein module $S(\partial T)$ instead of a local operator.

Let us consider a diagram of a string link with a positive quadruple crossing *quad* and let p be, e.g., the positive triple crossing associated

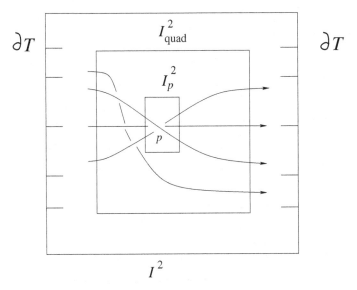

Figure 6.4: Increasing cubes.

with the adjacent stratum P_1. We consider three cubes: $I_p^2 \times I \subset$ $I_{\text{quad}}^2 \times I \subset I^2 \times I$. Here, I_p^2 is a small square which contains the triple point $pr(p)$ and I_{quad}^2 is a slightly bigger square which contains the quadruple point $pr(\text{quad})$. We illustrate this in Fig. 6.4. Let us replace the triple crossing p in $I_p^2 \times I$ by some standard element, say $L(p)$, in the skein module $S(\text{3-braids})$. Gluing to $L(p)$, the rest of the tangle T outside $I_p^2 \times I$ gives an element, say $T(p)$, in the skein module $S(\partial T)$. A triple crossing in an oriented isotopy corresponds to a Reidemeister move of type III (see, e.g., Ref. [9] for the definition of Reidemeister moves). We use the standard notations for braid groups (see, e.g., Ref. [1]).

Definition 6.3 *The Reidemeister move* $\sigma_1\sigma_2\sigma_1 \to \sigma_2\sigma_1\sigma_2$ *has sign* $= +1$ *and its inverse has sign* $= -1$.

Let m be the oriented meridian of the positive quadruple crossing. Our tetrahedron equation in $S(\partial T)$ is now

$$\sum_{p\in m} \text{sign}(p)T(p) = 0 \qquad (6.1)$$

where the sum is over all eight triple crossings in m.

It follows immediately from the definition of a skein module and from the fact that the tangle T is the same for all eight triple crossings outside $I_{\mathrm{quad}}^2 \times I$ that Equation (6.1) reduces to the local tetrahedron Equation (6.2) in the skein module S(4-braids):

$$\sum_{p \in m} \mathrm{sign}(p) L(p) = 0. \tag{6.2}$$

(Of course, we have to add here to each tangle in $L(p)$ the corresponding fourth strand in $I_{\mathrm{quad}}^2 \times I$, but we denote it still by $L(p)$.)

Proposition 6.1 *Equation (6.2) has a unique solution (up to a common factor), which is* $L(p) = \sigma_2(p) - \sigma_1(p)$.

We call this solution the *solution of constant weight* (the name will become clear later).

Proof. We prove only the existence. The unicity is less important and we left the proof to the reader. In general, for each permutation 3-braid β, we will write $\beta(p)$ for the element in the skein module, which is obtained by replacing the triple crossing by β. One has to use that the skein module of 3-braids is generated by the permutation braids 1, σ_1, σ_2, $\sigma_1\sigma_2$, $\sigma_2\sigma_1$ and $\sigma_1\sigma_2\sigma_1$. The latter permutation braid is not interesting because it is always just the original tangle. We call σ_1, σ_2 the *partial smoothings with one crossing* and $\sigma_1\sigma_2$, $\sigma_2\sigma_1$ the *partial smoothings with two crossings*. Equation (6.2) reduces now to a finite number of equations which have a single solution.

The signs of the triple crossings in the meridian are shown in Fig. 3.1 too.

First of all, we observe that for every basis element of the skein module of 3-braids, the contribution of the stratum P_2 cancels out with that of the stratum $\bar{P}_2$. The same is true for P_3 and $\bar{P}_3$. This comes from the fact that the signs are opposite but the resulting tangles are isotopic (there is just one branch which moves over or under everything else). Hence, we are left only with the strata P_1, P_4, $\bar{P}_1$, $\bar{P}_4$.

The substitution of the solution of constant weight into Equation (6.2) now gives:

$$-\sigma_1(\sigma_3 - \sigma_2)\sigma_1\sigma_2 + \sigma_2\sigma_1(\sigma_3 - \sigma_2)\sigma_1$$
$$+ \sigma_2\sigma_3(\sigma_2 - \sigma_1)\sigma_3 - \sigma_3(\sigma_2 - \sigma_1)\sigma_3\sigma_2$$

$$= (\sigma_2\sigma_1 + z\sigma_1\sigma_2\sigma_1 - \sigma_3\sigma_2 - z\sigma_1\sigma_3\sigma_2)$$
$$- (\sigma_1\sigma_2 + z\sigma_1\sigma_2\sigma_1 - \sigma_2\sigma_3 - z\sigma_2\sigma_1\sigma_3)$$
$$- (\sigma_2\sigma_1 + z\sigma_2\sigma_1\sigma_3 - \sigma_3\sigma_2 - z\sigma_3\sigma_2\sigma_3)$$
$$+ (\sigma_1\sigma_2 + z\sigma_1\sigma_3\sigma_2 - \sigma_2\sigma_3 - z\sigma_2\sigma_3\sigma_2) = 0. \qquad \square$$

In order to construct a 1-cocycle from a solution of the tetrahedron equation, we have to solve now in addition the *cube equations*.

It was only in spring 2012 that we have finally understood the complicated combinatorics which is needed in order to find a solution of the tetrahedron equation which uses the point at infinity and which uses exactly one of the only non-symmetric (under *flip*) partial smoothings $\sigma_1\sigma_2$ and $\sigma_2\sigma_1$.

Obviously, the contribution of $\sigma_2\sigma_1$ from $-P_2$ cancels out with that from $\bar{P}_2$ as well as the contribution from P_3 with that from $-\bar{P}_3$. Let us make a table of the contributions to $S(4-braids)$ of the partial smoothing $\sigma_2\sigma_1$ from the remaining four adjacent strata to a positive quadruple crossing together with the contribution $zL(p)$ of P_3. It is convenient to give them as planar pictures in Fig. 6.5 using the fact that all crossings are positive. Looking at Fig. 6.5, we make an astonishing discovery: The contribution of $\sigma_2\sigma_1$ cancels out in P_4 with that from $-\bar{P}_4$. Its contribution in $-P_1$ and $\bar{P}_1$ would cancel out with the contribution of $zL(p)$ in P_3 and $-\bar{P}_3$ if the latter triple crossings would count with some coefficient W such that $W(P_3) = W(\bar{P}_3) - 1$. (Of course, we have a symmetric result for $\sigma_1\sigma_2$ by interchanging P_4 with P_1 and P_3 with P_2.)

So, we are looking for an integer $W(p)$ which is defined for each diagram of a long knot with a positive triple crossing p and which has the following properties:

- $W(p)$ is an *isotopy invariant for the Reidemeister move* p, i.e., $W(p)$ is invariant under any isotopy of the rest of the tangle outside $I_p^2 \times I$. Note that $I_p^2 \times I$ is not a small cube around the triple crossing but its height is the height of the whole cube. In particular, no branch is allowed to move over or under the triple crossing;
- $W(p)$ depends non-trivially on the point at infinity;
- $W(P_3) = W(\bar{P}_3) - 1$ if and only if the triple crossing in P_1 (and hence in $\bar{P}_1$) is of a certain global type;
- $W(P_2) = W(\bar{P}_2)$ (this implies in fact the scan-property);
- $W(P_1) = W(\bar{P}_1) = W(P_4) = W(\bar{P}_4)$.

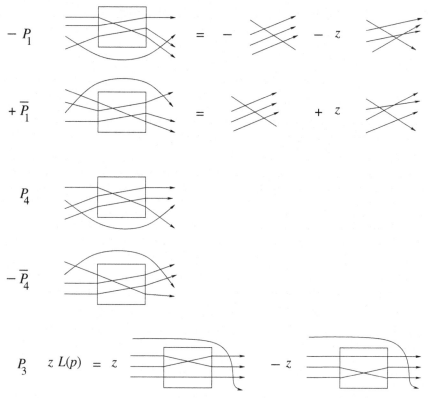

Figure 6.5:   The remaining partial smoothings $\sigma_2\sigma_1$ for the meridian of a positive quadruple crossing.

Assume that we have found such a $W(p)$. We define then

$$R(m) = \sum_{p\in m} \mathrm{sign}(p)\sigma_2\sigma_1(p) + \sum_{p\in m} \mathrm{sign}(p)zW(p)(\sigma_2(p) - \sigma_1(p))$$

where the first sum is only over certain global types of triple crossings (compare Fig. 6.6). It follows from the above considerations that $R(m) = 0$ and consequently the 1-cochain R is a new solution of the tetrahedron Equation (6.2).

It turns out that $W(p)$ does exist. We call it a *weight for the partial smoothing*. It is a finite type invariant of degree 1 for long knots relative to the triple crossing p and it is convenient to define it as a Gauss diagram invariant (for the definition of Gauss diagram

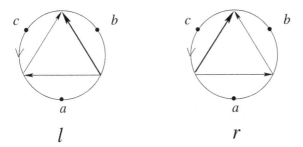

Figure 6.6:   The six global types of triple crossings.

invariants, see Refs. [15,40]). However, it is only invariant under regular isotopy and the corresponding 1-cocycle behaves uncontrollable under Reidemeister moves of type I.

So, let us recall the situation:

- there is a solution of the tetrahedron equation of *constant* weight, i.e., the partial smoothing $\sigma_2\sigma_1(p)$ enters with coefficient 0 and $L(p)$ enters with a coefficient of degree 0 (non-zero constant);
- there is a solution of the tetrahedron equation of *linear* weight, i.e., $\sigma_2\sigma_1(p)$ enters with a coefficient of degree 0 (non-zero constant) and $L(p)$ enters with a coefficient of degree 1.

The solution with linear weight leads to our 1-cocycle $R_{\text{reg}}^{(1)}$. The grading, mentioned in the Introduction, comes from the surprising fact that *the new contributing to $W(p)$ crossing in $\bar{P}_3$ with respect to P_3 is always identical to the crossing between the highest and the middle branch in the triple crossings P_1 and $\bar{P}_1$*. This grading is very useful for $R_{\text{reg}}^{(1)}(\text{scan}(T))$.

But these two solutions do not lead to a 1-cocycle which represents a non-trivial cohomology class in the special case of long knots, and moreover, the grading is trivial in this case. It turns out that we have to go one step further.

- There is a solution of the tetrahedron equation of *quadratic* weight, i.e., $\sigma_2\sigma_1(p)$ enters with a coefficient of degree 1 and $L(p)$ enters with a coefficient of degree 2.

Note that the weights $W(p)$ will depend on the whole long knot K and not only on the part of it in $I_{\text{quad}}^2 \times I$. Moreover, they depend crucially on the point at infinity and they cannot be defined for compact knots in S^3. Therefore, we have to consider 24 different positive tetrahedron equations, corresponding to the six different abstract closures of the four lines in $I^2_{\text{quad}} \times I$ to a circle and to the four different choices of the point at infinity in each of the six cases. $R^{(1)}$ and $R^{(1)}_{\text{reg}}$ are common solutions of all 24 tetrahedron equations and we call them therefore *solutions of a global positive tetrahedron equation*.

In order to define a 1-cocycle which represents a non-trivial cohomology class for long knots, one has to break the symmetry in a very strong way: *the partial smoothings $\sigma_2\sigma_1(p)$ and $\sigma_2(p) - \sigma_1(p)$ both enter with coefficients which depend on the whole long knot K but the partial smoothing $\sigma_1\sigma_2(p)$ doesn't enter at all.* For solutions with this property, all 24 tetrahedron equations become independent (there aren't any symmetries and the system becomes very over determined)! Their solutions give rise (after in addition the solving of the corresponding cube equations) to our 1-cocycles $R^{(1)}_{\text{reg}}$ and $R^{(1)}$. Moreover, $R^{(1)}$ vanishes for the meridians of a degenerate cusp and of the transverse intersections of arbitrary strata of codimension 1. It follows from the list of singularities of codimension two for projections of knots into the plan that $R^{(1)}$ is a 1-cocycle for all isotopies in the complement of the strata corresponding to a cusp with a transverse branch. We have succeeded to complete $R^{(1)}$ with a finite type Gauss diagram 1-cocycle $V^{(1)}$ of degree 3 (which has also the scan-property) in order to make it zero on the meridians of these remaining strata of codimension two. The result is our 1-cocycle $\bar{R}^{(1)}$ (compare the important Remark 6.11 in Section 6.10).

Remark 6.2 *We could make both $R^{(1)}$ and $V^{(1)}$ 1-cocycles in the whole M_T by "symmetrising" them (i.e., by considering linear combinations with "dual" 1-cocycles obtained by applying flip to the configurations or by using "heads" instead of "foots" of arrows, compare Sections 6.4 and 6.5). But it seems likely that each of them would represent the trivial cohomology class.*

We have carried out our construction also for the 2-variable Kauffman polynomial F instead of the HOMFLYPT polynomial. The skein relations for the 2-variable Kauffman invariant are shown in Fig. 6.7 (see Ref. [31]).

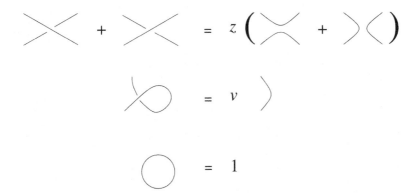

Figure 6.7: Skein relations for the Kauffman invariant.

Remark 6.3 *Note that the solution of the global positive tetrahedron equation in the case of the HOMFLYPT polynomial can be expressed by using only partial smoothings with* two *crossings:*

$$R(m) = \sum_{p \in m} \mathrm{sign}(p)\sigma_2\sigma_1(p) + \sum_{p \in m} \mathrm{sign}(p)W(p)(\sigma_2^2(p) - \sigma_1^2(p))$$

where the first sum is only over certain global types of triple crossings. Amazingly, exactly the same formula is also a solution in the case of the Kauffman polynomial!

However, we haven't solved the cube equations in this case. They are very complicated.

Surprisingly, there is a second solution of the global positive tetrahedron equation in Kauffman's case (which does not exist in the HOMFLYPT case). We give it in Fig. 6.8. The weights and the global types of triple crossings are exactly the same as in the previous solutions.

We solve the cube equations for this solution with coefficients in $\mathbb{Z}/2\mathbb{Z}$. The results are two other quantum 1-cocycles, called $R_{F,\mathrm{reg}}^{(1)}$ and $R_F^{(1)}$ which also have the scan-property. We can complete $R_F^{(1)}$ with exactly the same 1-cocycle $V^{(1)}$ as in the HOMFLYPT case:

$$\bar{R}_F^{(1)} = R_F^{(1)} + F_T V^{(1)}.$$

A calculation by hand now gives

$$\bar{R}_F^{(1)}(rot(3_1^+)) = F_{3_1^+} \bmod 2.$$

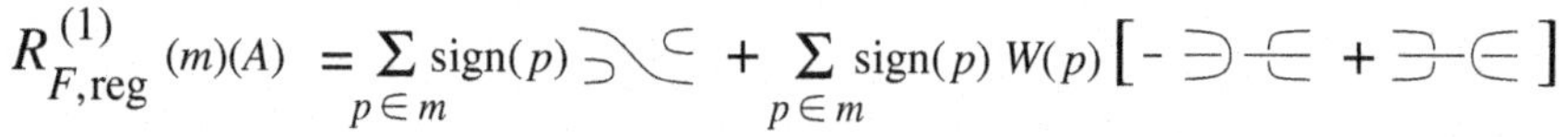

$$R^{(1)}_{F,\text{reg}}(m)(A) = \sum_{p\,\in\,m} \text{sign}(p)\; \raisebox{-0.3em}{\includegraphics{}}\; + \sum_{p\,\in\,m} \text{sign}(p)\, W(p) \left[- \raisebox{-0.3em}{\includegraphics{}} + \raisebox{-0.3em}{\includegraphics{}} \right]$$

Figure 6.8: Surprising solution of the positive tetrahedron equation in Kauffman's case.

Consequently, $\bar{R}^{(1)}_F$ also represents a non-trivial cohomology class in the topological moduli spaces M_T and it has the scan-property too.

It is well known that the coloured HOMFLYPT polynomial, i.e., the knot invariant associated with representations of sl_N, can be expressed as a linear combination of usual HOMFLYPT polynomials applied to cables. Hence, we define the corresponding 1-cocycles just as linear combinations of the 1-cocycles $\bar{R}^{(1)}$ applied to the cables $Cab_n(K)$ of long knots K. (We define a substitute for $rot(Cab_n(K))$ in Remark 6.10 in Section 6.11.)

6.2   The Topological Moduli Space of Diagrams of String Links and Its Stratification

We work in the smooth category.

We fix an orthogonal projection $pr : I^3 \to I^2$. Let T be a string link. T is called *regular* if $pr : T \to I^2$ is an immersion. Let T be a regular string link and let M_T be the (infinite dimensional) space of all string links isotopic to T. We denote by M_T^{reg} the sub-space of all string links which are regularly isotopic to T, i.e., all string links in the isotopy are regular. The infinite dimensional spaces M_T and M_T^{reg} have a natural *stratification with respect to* pr:

$$M_T = \Sigma^{(0)} \cup \Sigma^{(1)} \cup \Sigma^{(2)} \cup \Sigma^{(3)} \cup \Sigma^{(4)} \dots$$

Here, $\Sigma^{(i)}$ denotes the union of all strata of codimension i.

The strata of codimension 0 correspond to the usual generic *diagrams of tangles*, i.e., all singularities in the projection are ordinary double points. So, our *discriminant* is the complement of $\Sigma^{(0)}$ in M_T.

The three types of strata of codimension 1 correspond to the *Reidemeister moves*, i.e., non-generic diagrams which have exactly one ordinary triple point, denoted by $\Sigma^{(1)}_{\text{tri}}$, or one ordinary self-tangency, denoted by $\Sigma^{(1)}_{\text{tan}}$, or one ordinary cusp, denoted by $\Sigma^{(1)}_{\text{cusp}}$, in the projection *pr*. We call the triple point together with the under–over information (i.e., its embedded resolution in I^3 given by the tangle T) a *triple crossing*.

There are exactly six types of strata of codimension 2. They correspond to non-generic diagrams which have exactly either

- one ordinary quadruple point, denoted by $\Sigma_{\text{quad}}^{(2)}$;
- one ordinary self-tangency with a transverse branch passing through the tangent point, denoted by $\Sigma_{\text{trans-self}}^{(2)}$;
- one self-tangency in an ordinary flex $(x = y^3)$, denoted by $\Sigma_{\text{self-flex}}^{(2)}$;
- two singularities of codimension 1 in disjoint small discs (this corresponds to the transverse intersection of two strata from $\Sigma^{(1)}$);
- one ordinary cusp with a transverse branch passing through the cusp, denoted by $\Sigma_{\text{trans-cusp}}^{(2)}$; or
- one degenerate cusp, locally given by $x^2 = y^5$, denoted by $\Sigma_{\text{cusp-deg}}^{(2)}$.

We show these strata in Fig. 1.1. (Compare also Ref. [19], which contains all the necessary preparations from singularity theory.)

The strata which we need from $\Sigma^{(3)}$ will be described later.

The above list of singularities of codimension two can be seen as an analogue for one parameter families of diagrams of the Reidemeister theorem (which gives the list of singularities of codimension one).

The procedure in the combinatorial approach is always the following:

(1) associating something calculable (e.g., an element in some algebra) to a point in $\Sigma^{(0)}$;
(2) proving the invariance under passing strata from $\Sigma^{(1)}$;
(3) simplifying this proof by using strata from $\Sigma^{(2)}$.

A generic isotopy of string links is an arc in M_T and the intersection with $\Sigma^{(i)}$ is empty for $i > 1$.

Invariants which are constructed in this way are consequently 0-cocycles for M_T with values in some algebra.

The main tool (1) is quantum invariant, i.e., elements I of a *skein module* coming from irreducible representations of quantum groups

$U_q(g)$ (here, $U_q(g)$ is the quantum enveloping algebra of a complex simple Lie algebra g at a non-zero complex number q which is not a root of unity) or more generally of ribbon Hopf algebras (see, e.g., Refs. [28,44]).

(2) reduces essentially to the proof of invariance of I under passing a stratum of $\Sigma^{(1)}$ which corresponds to a non-generic diagram which has exactly one ordinary positive triple crossing. In the case of quantum invariants, this reduces to prove that quantum invariants are solutions of the *Yang–Baxter equation*.

(3) is not explicitly mentioned (or only in a sloppy way) in many papers. We will see later that there are exactly two global (without taking into account the point at infinity) and eight local types of triple crossings. The Yang–Baxter equation does not depend on the global type of the triple crossing. Hence, we stay with exactly eight different Yang–Baxter equations. Different local types of triple crossings come together in points of $\Sigma^{(2)}_{\text{trans}-\text{self}}$. We make a graph Γ in the following way: the vertices correspond to the different local types of triple crossings. We connect two vertices by an edge if and only if the corresponding strata of triple crossings are adjacent to a stratum of $\Sigma^{(2)}_{\text{trans}-\text{self}}$. One easily sees that the resulting graph is the 1-skeleton of the 3-dimensional cube I^3 (compare Section 6.6). In particular, it is connected. Studying the normal discs to $\Sigma^{(2)}_{\text{trans}-\text{self}}$ in M_T, one shows that if I is invariant under passing all $\Sigma^{(1)}_{\text{tan}}$ and just *one* local type of a stratum $\Sigma^{(1)}_{\text{tri}}$, then it is invariant under passing all types of triple crossings because Γ is connected. Hence, the use of certain strata of $\Sigma^{(2)}$ allows us to reduce the eight Yang–Baxter equations to a single one.

We call the resulting 0-cocycles *quantum 0-cocycles*. These are the usual well-known quantum invariants. All quantum invariants verify *skein relations* (see Figs. 6.1 and 6.7) and hence give rise to *skein modules*, i.e., formal linear combinations over the corresponding ring of coefficients of tangles (with the same oriented boundary) modulo the skein relations. Skein modules were introduced by Turaev [44] and Przytycki [41]. The *multiplication of tangles* is defined by their composition (if possible), see Fig. 6.9. The main property of quantum invariants (and which follows directly from their construction) is their *multiplicativity*: let T_1 and T_2 be tangles which can be composed

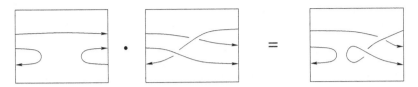

Figure 6.9:   Multiplication of tangles.

and let $I(T_i)$ be the corresponding quantum invariants in the skein modules. Then, $I(T_1T_2) = I(T_1)I(T_2)$, where we have to decompose, by using the skein relations again, the product of the generators of the skein modules for ∂T_1 with those for ∂T_2 into a linear combination of generators for the skein module of $\partial(T_1T_2)$. In particular, if the n-tangle T_2 consists of n trivial arcs with just a local knot K tied into one of the arcs, then $I(T_1T_2) = I(T_1)I(K)$. Consequently, a Reidemeister move of type I (i.e., a connected sum with a small diagram of the unknot which has just one crossing) changes $I(T_1)$ only by some standard factor.

The most famous quantum invariants are the HOMFLYPT polynomial P, the 2-variable Kauffman polynomial F and Kuperberg's polynomial I^{g_2} and their generalisations for tangles.

6.3   The Strategy of the Construction and Notations

The construction of the 1-cocycles is very long and difficult. But once we have them, their applications will be much simpler.

The main idea consists in using strata $\Sigma^{(i)}$ of the discriminant of codimensions $i = 0, 1, 2, 3$ in order to construct quantum 1-cocycles in a similar way as the strata of codimensions 0, 1 and 2 were used in order to construct quantum 0-cocycles.

First of all, it turns out that starting from 1-cocycles we have to replace in general isotopy by regular isotopy. This comes from the fact that the invariants which we will construct are no longer multiplicative and hence the moment (in a family of diagrams) and the place (in the diagram) of a Reidemeister move of type I become important. Fortunately, this is not restrictive in order to distinguish tangles. It is well known that two (ordered oriented) tangles are regularly isotopic if and only if they are isotopic and the corresponding components share the same *Whitney index* $n(T)$ and the same *writhe*

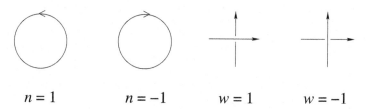

$$n = 1 \qquad n = -1 \qquad w = 1 \qquad w = -1$$

Figure 6.10:   Whitney index and writhe.

$w(T)$ (see, e.g., Ref. [15]). Here, as usual, the Whitney index is the algebraic number of oriented circles in the plane after smoothing all double points of a component with respect to the orientation and the writhe is the algebraic number of all crossings of a component. We use the standard conventions shown in Fig. 6.10. Hence, by adding if necessary small curls to components we can restrict ourself to regular isotopy.

We choose a quantum 0-cocycle, e.g., the regular HOMFLYPT invariant in the regular HOMFLYPT skein module $S(\partial T)$. We fix moreover an abstract closure σ of T such that it becomes an oriented circle (just a part of the circle is embedded as T in I^3) and we fix a *point at infinity* in ∂T.

(1) A generic arc s in M_T^{reg} intersects $\Sigma_{\text{tri}}^{(1)}$ and $\Sigma_{\text{tan}}^{(1)}$ transversally in a finite number of points. To each such intersection point we associate an element in the HOMFLYPT skein module $S(\partial T)$. We sum up with signs the contributions over all intersection points in s and obtain an element, say $R(s)$, in $S(\partial T)$.

(2) We use now the strata from $\Sigma^{(2)}$ to show the invariance of $R(s)$ under all generic homotopies of s in M_T^{reg} which fix the end points of s. The contributions to $R(s)$ come from discrete points in s and hence they survive under generic Morse modifications of s. It follows that $R(s)$ is a 1-cocycle for M_T^{reg}.

(3) Again, we use some strata from $\Sigma^{(3)}$ in order to simplify the proof that $R(s)$ is a 1-cocycle.

In (1), we do something new for triple crossings and for self-tangencies, but we treat the usual crossings as previously in the construction of 0-cocycles. This is forced by those strata in $\Sigma^{(2)}$ which consist of the transverse intersections of two strata in $\Sigma^{(1)}$.

In (2), we have to study normal 2-discs for the strata in $\Sigma^{(2)}$. For each type of stratum in $\Sigma^{(2)}$, we have to show that $R(m) = 0$ for the boundary m of the corresponding normal 2-disc. We call m a *meridian*. $\Sigma_{\text{quad}}^{(2)}$ is by fare the hardest case and we call the corresponding equation $R(m) = 0$ the *tetrahedron equation*. Indeed, as already mentioned in the Introduction, the quadruple crossing deforms into four different triple crossings (the vertices's of a tetrahedron) and the six involved ordinary crossings form the edges of the tetrahedron. (This is an analogue of the Yang–Baxter or triangle equation in the case of 0-cocycles.) It turns out that without considering the point at infinity there are six global types (corresponding to the *Gauss diagrams* but without the writhe of the quadruple crossings, see the end of the section) and 48 local types of quadruple crossings. At each point of $\Sigma_{\text{quad}}^{(2)}$, there are adjacent exactly eight strata of $\Sigma_{\text{tri}}^{(1)}$ (compare Fig. 3.2).

The edges of the graph $\Gamma = skl_1(I^3)$, which was constructed in the last section, correspond to the types of strata in $\Sigma_{\text{trans-self}}^{(2)}$. The solution of the tetrahedron equation tells us what is the contribution to R of a positive triple crossing (i.e., all three involved crossings are positive). The meridians of the strata from $\Sigma_{\text{trans-self}}^{(2)}$ give equations which allow us to determine the contributions of all other types of triple crossings as well as the contributions of self-tangencies. However, each loop in Γ could give an additional equation. Evidently, it suffices to consider the loops which are the boundaries of the 2-faces from $skl_2(I^3)$. In fact, they do give additional equations (see Sections 6.6 and 6.7) and force us to consider smoothings of self-tangencies too. We call all the equations which come from the meridians of $\Sigma_{\text{trans-self}}^{(2)}$ and from the loops in $\Gamma = skl_1(I^3)$ the *cube equations*. (Note that a loop in Γ is more general than a loop in M_T. For a loop in Γ, we come back to the same local type of a triple crossing but not necessarily to the same whole diagram of T.)

In (3), we consider the strata from $\Sigma^{(3)}$ which correspond to diagrams which have a degenerate quadruple crossing where exactly two branches have an ordinary self-tangency in the quadruple point, denoted by $\Sigma_{\text{trans-trans-self}}^{(3)}$ (see Fig. 6.11). (These strata are the analogue of the strata $\Sigma_{\text{trans-self}}^{(2)}$ in the construction of 0-cocycles.) Again, we form a graph with the local types of quadruple crossings

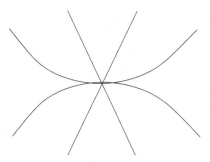

Figure 6.11:   A quadruple crossing with two tangential branches.

as vertices and the adjacent strata of $\Sigma^{(3)}_{\text{trans-trans-self}}$ as edges. One easily sees that the resulting graph Γ has exactly 48 vertices and that it is again connected. We don't need to study the unfolding of $\Sigma^{(3)}_{\text{trans-trans-self}}$ in detail. It is clear that each meridional 2-sphere for $\Sigma^{(3)}_{\text{trans-trans-self}}$ intersects $\Sigma^{(2)}$ transversally in a finite number of points, namely, exactly in two strata from $\Sigma^{(2)}_{\text{quad}}$ and in lots of strata from $\Sigma^{(2)}_{\text{trans-self}}$ and from $\Sigma^{(1)} \cap \Sigma^{(1)}$, and there are no other intersections with strata of codimension 2. If we know now that $R(m) = 0$ for the meridian m of one of the quadruple crossings, that $R(m) = 0$ for all meridians of $\Sigma^{(2)}_{\text{trans-self}}$ (i.e., R satisfies the cube equations) and for all meridians of $\Sigma^{(1)} \cap \Sigma^{(1)}$, then $R(m) = 0$ for the other quadruple crossing too. It follows that for each fixed global type (see Fig. 3.3) the 48 tetrahedron equations reduce to a single one, which we call the *positive tetrahedron equation* (compare the Introduction). This phenomenon is analogue to that for the Yang–Baxter or triangle equation, which was explained in the previous section. We now need a solution of the positive tetrahedron equation which depends on a point at infinity but which is a solution independent of the position of this point at infinity in the knot diagram.

Let T be a generic n-tangle in I^3. We assume that a half of ∂T is contained in $0 \times I^2$ and the other half in $1 \times I^2$, and we fix an abstract closure σ of T such that it becomes an oriented circle (compare, e.g., Example 6.1 in Section 6.8).

The sign of a positive Reidemeister III move was already defined in the Introduction. The signs of the remaining seven Reidemeister III moves are determined by the meridians of the strata $\Sigma^{(2)}_{\text{trans}-\text{self}}$. It is convenient to equip the strata of $\Sigma^{(1)}_{\text{tri}}$ with a co-orientation. The sign is then defined as the local intersection index, which is $+1$ if the oriented isotopy intersects $\Sigma^{(1)}$ transversally from the negative side to the positive side. We give the local types of Reidemeister moves of type III together with the co-orientation in Fig. 2.2. The $+$ or $-$ sign here denotes the side of the complement of $\Sigma^{(1)}_{\text{tri}}$ with respect to the co-orientation. An easy verification shows that they satisfy the cube equations, i.e., the signs are consistent for the boundary of each 2-cell in $skl_2(I^3)$. The six local types $1, 3, 4, 5, 7, 8$ can appear for braids and are therefore called *braid-like*. The remaining two types $2, 6$ are called *star-like*. In the star-like case, we denote by *mid* the boundary point which corresponds to the ingoing middle branch.

Definition 6.4 *The sign of a Reidemeister II move and of a Reidemeister I move is shown in Fig. 6.12.*

Note that these signs depend only on the underlying planar curve in contrast to the signs for triple crossings.

To each Reidemeister move of type III corresponds a diagram with a *triple crossing p*: three branches of the tangle (the highest, middle and lowest with respect to the projection pr) have a common point in the projection into the plane. A small perturbation of the triple crossing leads to an ordinary diagram with three crossings near $pr(p)$.

Definition 6.5 *We call the crossing between the highest and the lowest branch of the triple crossing p the distinguished crossing of p and we denote it by d. The crossing between the highest branch and the middle branch is denoted by hm and that of the middle branch with the lowest is denoted by ml. Smoothing the distinguished crossing with respect to the orientation splits $T \cup \sigma$ into two oriented and ordered circles. We call D^+_d the component which goes from the undercross*

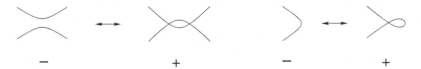

Figure 6.12:   Coorientation for Reidemeister moves of types II and I.

to the overcross at d and by D_d^- the remaining component (compare Fig. 1.4).

In a Reidemeister move of type II, both new crossings are considered as distinguished and denoted by d.

For the definition of the weights $W(p)$, we use Gauss diagram formulas (see Refs. [15,40]). A *Gauss diagram of* K is an oriented circle with oriented chords and a marked point. There exists an orientation preserving diffeomorphism from the oriented line to the oriented knot $T \cup \sigma$ such that each chord connects a pair of points which are mapped onto a crossing of $pr(T)$ and infinity is mapped to the marked point. The chords are oriented from the preimage of the undercrossing to the preimage of the overcrossing (here, we use the orientation of the I-factor in $I^2 \times I$). The circle of a Gauss diagram in the plan is always equipped with the counterclockwise orientation.

A *Gauss diagram formula* of degree k is an expression assigned to a knot diagram which is of the following form:

$$\sum \text{function(writhes of the crossings)}$$

where the sum is taken over all possible choices of k (unordered) different crossings in the knot diagram such that the chords arising from these crossings in the knot diagram of $T \cup \sigma$ build a given sub-diagram with given marking. The marked sub-diagrams are called *configurations*. If the function is the product of the writhes of the crossings in the configuration, then we will denote the sum shortly by the configuration itself.

A Gauss diagram formula which is invariant under regular isotopies of $T \cup \sigma$ is called a *Gauss diagram invariant*.

Let us consider the *Gauss diagram of a triple crossing p* but without taking into account the writhe of the crossings. In the oriented circle $T \cup \sigma$, we connect the preimages of the triple point $pr(p)$ by arrows which go from the undercross to the overcross and we obtain a triangle. The distinguished crossing d is always drawn by a thicker arrow.

There are exactly six different *global types* of triple crossings with a point at infinity. We give names to them and show them in Fig. 6.6. (Here, "r" indicates that the crossing between the middle and the lowest branch goes to the right and "l" indicates that it goes to the left.) Note that the involution *flip* exchanges the types r_a and l_a as well as the types r_b with l_c and r_c with l_b.

Figures 3.5–3.16 are derived from Figs. 3.1 and 3.3. They show the six global types of positive quadruple crossings where we give the names $1, 2, 3, 4$ to the four points at infinity. Each of the six crossings involved in a quadruple crossing gets the name of the two crossing lines.

For each of the adjacent eight strata of triple crossings, we show the Gauss diagrams of the six crossings with the names of the crossings which are not in the triangle (the names of the latter can easily be established from the figure too). These 12 figures are our main instrument in this chapter.

We make the following convention for the whole chapter: crossings which survive in isotopies (i.e., they are not involved in Reidemeister moves of types I or II) are identified. In particular, they get the same name.

Remark 6.4 *Our co-orientation for Reidemeister III moves is purely* local. *In Ref. [16], we have used a different co-orientation which is completely determined by the unoriented planar curves. We show it in Fig. 2.3. Let us call it the* global co-orientation. *To the three crossings of a Reidemeister III move, we associate chords in a circle which parametrises the knot. Note that for positive triple crossings of global type r, the local and the global co-orientation coincide and that for the global type l they are opposite. Comparing the global types I and III of positive quadruple points (compare Fig. 1.3), we see that there is no solution with constant weight of the global positive tetrahedron equation for the global co-orientation. Indeed, for the type I, we have the equation $-P_1 + P_4 + \bar{P}_1 - \bar{P}_4 = 0$ and for the type III we have $-P_1 - P_4 + \bar{P}_1 + \bar{P}_4 = 0$. Hence, the use of our local co-orientation for triple crossings is essential in our approach!*

6.4 Solution with Linear Weight of the Positive Global Tetrahedron Equation

We denote the point at infinity in the circle $\sigma \cup T$ by ∞.

Definition 6.6 *An ordinary crossing q in a diagram T with closure σ is of type 1 if $\infty \in D_q^+$ and is of type 0 otherwise.*

Let $T(p)$ be a generic diagram with a triple crossing or a self-tangency p (in other words, $T(p)$ is an interior point of $\bar{\Sigma}^{(1)}$) and let d be the distinguished crossing for p. In the case of a self-tangency, we identify the two distinguished crossings. *We assume that the distinguished crossing d is of type 0.*

Definition 6.7 *A crossing q of $T(p)$ is called an f-crossing if q is of type 1 and the under cross of q is in the oriented arc from ∞ to the over cross of d in $\sigma \cup T$.*

In other words, the "foot" of q in the Gauss diagram is in the sub-arc of the circle which goes from ∞ to the head of d, and the head of q is in the sub-arc which goes from the foot of q to ∞. We illustrate this in Figs. 6.13 and 6.14. (The letter "f" stands for "foot" or "Fuß" and not for "fiedler.") Note that we make essential use of the fact that no crossing can ever move over the point at infinity.

Definition 6.8 *The* linear weight $W(p)$ *is the Gauss diagram invariant shown in Fig. 6.13. In other words, it is the sum of the writhes of all f-crossings with respect to p in a given diagram $T(p)$. Note that if p is a self-tangency, then the degenerate second configuration in Fig. 6.13 cannot appear.*

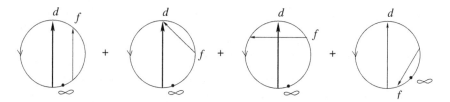

Figure 6.13:   The f-crossings.

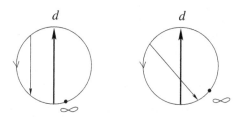

Figure 6.14:   Crossings of type 1 which are not f-crossings.

Definition 6.9 *Let q be an f-crossing. The grading $A = \partial q$ of q is defined by $\partial q = D_q^+ \cap \partial T$.*

The grading ∂q is a cyclically ordered sub-set of ∂T with alternating signs and which contains ∞ (compare the Introduction).

Lemma 6.1 *The graded $W(p)$ is an isotopy invariant for each Reidemeister move of type II or III, i.e., $W(p)$ is invariant under any isotopy of the rest of the tangle outside of $I_p^2 \times I$ (compare the Introduction).*

In other words, $W(p)$ doesn't change under a homotopy of an arc which passes through $\Sigma^{(1)} \cap \Sigma^{(1)}$ in M_T^{reg}.

Proof. This is obvious. One has only to observe that the two new crossings from a Reidemeister move of type II are either both not f-crossings or are both f-crossings. In the latter case, they have the same grading but they have different writhe. □

We denote by $W_A(p)$ the invariant $W(p)$ restricted to all f-crossings with a given grading A.

It is clear that $W_A(p)$ is an invariant of degree 1 for (T, p) for each fixed grading A (see, e.g., Refs. [15,40] for a proof that Gauss diagram invariants are of finite type).

Let us consider a global positive quadruple crossing with a fixed point at infinity (compare the Introduction and Section 6.3). The following lemma is of crucial importance.

Lemma 6.2 *The f-crossings of the eight adjacent strata of triple crossings have the following properties:*

- *the f-crossings in P_2 are identical with those in $\bar{P}_2$ (with our convention above);*
- *the f-crossings in P_1, $\bar{P}_1$, P_4 and $\bar{P}_4$ are all identical;*
- *the f-crossings in P_3 are either identical with those in $\bar{P}_3$ or there is exactly one new f-crossing in $\bar{P}_3$ with respect to P_3. In the latter case, the new crossing is always exactly the crossing hm in P_1 and $\bar{P}_1$;*
- *the new f-crossing in $\bar{P}_3$ appears if and only if P_1 (and hence also $\bar{P}_1$) is one of the two global types shown in Fig. 6.15.*

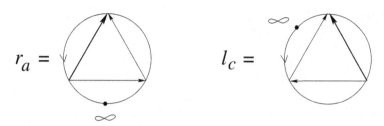

Figure 6.15:   The global types for which a new f-crossing in $\bar{P}_3$ appears.

Proof. We have checked the assertions of the lemma in all twenty-four cases (denoted by the global type of the quadruple crossing together with the point at infinity) using Figs. 3.5–3.16. Note that the crossing hm in P_1 and $\bar{P}_1$ is always the crossing 34 and that d is of type 0 if ∞ is on the right-hand side of it in the figures. We consider just some examples. In particular, we show that it is necessary to add the "degenerate" configuration in Fig. 6.13 and we left the rest of the verification to the reader.

The f-crossings are as follows:

Case I_1: None at all.

Case I_2: None at all.

Case I_3: In $P_1, \bar{P}_1$: 34 (both are the degenerate case). In $P_4, \bar{P}_4$: 34 (the third configuration and the first configuration in Fig. 6.13). In $P_2, \bar{P}_2$: 34. In P_3: non. In $\bar{P}_3$: 34.

Case I_4: In $P_1, \bar{P}_1, P_4, \bar{P}_4$: 34, 24, 23. In $P_2, \bar{P}_2$: non. In P_3: 23, 24. In $\bar{P}_3$: 23, 24, 34.

Case VI_3: In P_2: 12, 13, 14. In $\bar{P}_2$: 12, 13, 14 (12 shows that the fourth configuration in Fig. 6.13 is necessary too).

Case VI_1: In P_3: None. In $\bar{P}_3$: 34 (shows that the second configuration in Fig. 6.15 is necessary too).   □

Let γ be an oriented generic arc in M_T^{reg} which intersects $\Sigma^{(1)}$ only in positive triple crossings and let A be a fixed element of the gradings.

Definition 6.10 *The evaluation of the 1-cochain $R_{\text{reg}}^{(1)}$ with grading A on γ is defined by*

$$R_{\text{reg}}^{(1)}(A)(\gamma) = \sum_{p \in \gamma} \text{sign}(p)\sigma_2\sigma_1(p)$$

$$+ \sum_{p \in \gamma} \text{sign}(p)z W_A(p)(\sigma_2(p) - \sigma_1(p))$$

where the first sum is over all triple crossings of the global types shown in Fig. 6.15 (i.e., the types l_c and r_a) and such that $\partial(hm) = A$. The second sum is over all triple crossings p which have a distinguished crossing d of type 0 (i.e., the types r_a, r_b and l_b).

(Remember that, e.g., $\sigma_2\sigma_1(p)$ denotes the element in the HOM-FLYPT skein module $S(\partial T)$ which is obtained by replacing the triple crossing p in the tangle T by the partial smoothing $\sigma_2\sigma_1$, compare the Introduction.)

Note that a triple crossing of the first type in Fig. 6.15 could also contribute to the second sum in $R_{\text{reg}}^{(1)}(A)(\gamma)$ but a triple crossing of the second type in Fig. 6.15 can't because d is of type 1.

Proposition 6.2 *Let m be the meridian for a positive quadruple crossing. Then, $R_{\text{reg}}^{(1)}(A)(m) = 0$ for each grading A.*

Proof. We have to consider the contributions of the eight strata $P_1, \ldots, \bar{P}_4$.

Case 1: The f-crossings of grading A are identical in P_3 and $\bar{P}_3$.

It follows from Lemma 6.2 that either the new f-crossing in $\bar{P}_3$ is not of grading A or there is no new f-crossing in $\bar{P}_3$ at all. Moreover, in the first case, the crossing hm in P_1 is not of grading A and hence P_1 and $\bar{P}_1$ do not contribute to the first sum in $R_{\text{reg}}^{(1)}(A)(m)$. In the second case, the triple crossing P_1 (and consequently $\bar{P}_1$ too) is not of the type required in Fig. 6.15 and hence again does not contribute to the first sum.

It follows from Lemma 6.2 that $W_A(P_2) = W_A(\bar{P}_2)$ as well as $W_A(P_3) = W_A(\bar{P}_3)$. Moreover, each partial smoothing of P_2 (respectively, P_3) is regularly isotopic to the same partial smoothing of $\bar{P}_2$

(respectively, $\bar{P}_3$). (This follows from the fact that the fourth branch just moves over or under everything.) But P_i and $\bar{P}_i$ always enter with different signs. It follows that all contributions of these strata to $R_{\text{reg}}^{(1)}(A)(m)$ cancel out.

The contributions of P_4 and $\bar{P}_4$ to the first sum in $R_{\text{reg}}^{(1)}(A)(m)$ cancel out as was shown in Fig. 6.5.

It follows again from Lemma 6.2 that P_1, $\bar{P}_1$, P_4, $\bar{P}_4$ always share the same W_A. Consequently, the contribution of these strata to the second sum in $R_{\text{reg}}^{(1)}(A)(m)$ is just a multiple of the solution with constant weight (compare the Introduction) and hence, it is 0 by Proposition 6.1.

Case 2: The f-crossings of grading A are not identical in P_3 and $\bar{P}_3$.

In this case, the new f-crossing in $\bar{P}_3$ is of grading A. It follows from Lemma 6.2 that the crossing hm in P_1 (and hence in $\bar{P}_1$ too) is of grading A. $W_1(\bar{P}_3) = W_A(P_3) + 1$ because all crossings are positive. Using Fig. 3.1, we find the contributions of P_1, $\bar{P}_1$, P_3, $\bar{P}_3$ to $R_{\text{reg}}^{(1)}(A)(m)$ and the calculation is already given in Fig. 6.5. The rest of the arguments are the same as in Case 1.   □

We have proven that $R_{\text{reg}}^{(1)}(A)$ is a solution of the positive global tetrahedron equation.

6.5 Solution with Quadratic Weight of the Positive Global Tetrahedron Equation

We start with recalling the Gauss diagram formula of Polyak and Viro for the Vassiliev invariant $v_2(K)$ (see Ref. [40]). In fact, there are two formulas shown in Fig. 6.16. These formulas are for knots in the 3-sphere and the marked point can be chosen arbitrary on the knot. In particular, in the case of long knots, we chose of course ∞ as the marked point.

Let $T(p)$ be a generic diagram with a triple crossing or a self-tangency p (in other words, $T(p)$ is an interior point of $\Sigma^{(1)}$) and let d be the distinguished crossing for p. We assume that d is of type 0.

Definition 6.11 *The quadratic weight $W_2(p)$ is defined by the Gauss diagram formula shown in Fig. 6.16. Here, f denotes an f-crossing*

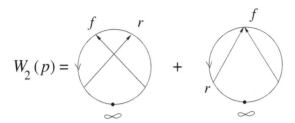

Figure 6.16: The quadratic weight W_2.

of p and r denotes an arbitrary crossing with a position as shown in the configurations.

(Remember that with our conventions the formula in Fig. 6.16 says that $W_2(p)$ is the sum of $w(f)w(r)$ over all couples of crossings (f, r) in $T(p)$ which form one of the configurations from Fig. 6.16.)

The second (degenerate) configuration in Fig. 6.16 of course cannot appear for a self-tangency p. Note that the crossing r is always of type 0.

The crossings r in Fig. 6.16 are called the *r-crossings*.

Lemma 6.3 $W_2(p)$ *is an isotopy invariant for each Reidemeister move of type I, II or III, i.e., $W_2(p)$ is invariant under any isotopy of the rest of the tangle outside $I_p^2 \times I$ (compare the Introduction).*

The lemma implies that as in the linear case $W_2(p)$ doesn't change under a homotopy of an arc which passes through $\Sigma^{(1)} \cap \Sigma^{(1)}$ but this time even in M_T. However, we have lost the grading.

Proof. It is obvious that $W_2(p)$ is invariant under Reidemeister moves of types I and II. The latter comes from the fact that both new crossings are simultaneously crossings of type f or r and that they have different writhe. As was explained in the Introduction, the graph Γ now implies that it is sufficient to prove the invariance of $W_2(p)$ only under positive Reidemeister moves of type III. There are two global types, and for each of them, there are three possibilities for the point at infinity. We give names $1, 2, 3$ to the crossings and a, b, c to the points at infinity, and we show the six cases in Figs. 6.17 and 6.18. Evidently, we have only to consider the mutual position of

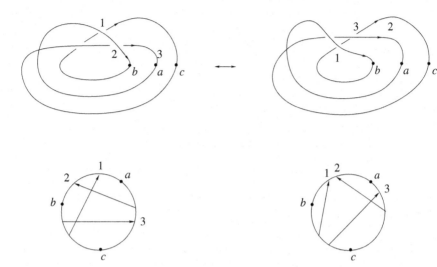

Figure 6.17: R III of type r.

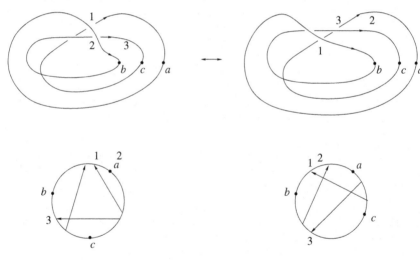

Figure 6.18: R III of type l.

the three crossings in the pictures because the contributions with all other crossings do not change.

r_a: There is only 3 which could be an f-crossing, but it does not contribute with another crossing in the picture to $W_2(p)$.

r_b: No r-crossing at all.

r_c: Only 2 could be an f-crossing. In that case, it contributes on the left-hand side exactly with the r-crossing 1 and on the right-hand side exactly with the r-crossing 3.

l_a: No f-crossing at all.

l_b: 1 and 2 could be f-crossings but none of them contributes with the crossing 3 to $W_2(p)$.

l_c: 1 and 3 could be f-crossings. But the foots of 1 and 3 are arbitrary close. Consequently, they can be f-crossings only simultaneously! In that case, 3 contributes with 2 on the left-hand side and 1 with 2 on the right-hand side.

Consequently, we have proven that $W_2(p)$ is invariant. But we have lost the grading in l_c. The f-crossings which contribute are not the same on the left-hand side and on the right-hand side, and we cannot guarantee that the crossings 1 and 3 have the same grading. □

We want to replace $W(p)$ by $W_2(p)$ for the partial smoothing $z(\sigma_2(p) - \sigma_1(p))$.

Let us denote by $W_2(p, f)$ the contribution of a given f-crossing f to $W_2(p)$.

In the case that the f-crossings in P_3 and $\bar{P}_3$ are not the same, we now get

$$W_2(\bar{P}_3) = W_2(P_3) + W_2(p, f)$$

where f is the new f-crossing in $\bar{P}_3$. But luckily the new f-crossing f is always just the crossing hm in P_1 (compare Lemma 6.2). This leads us to the following definition.

Definition 6.12 *Let p be a triple crossing of one of the types shown in Fig. 6.15. The linear weight $W_1(p)$ is defined as the sum of the writhes of the crossings r in $T(p)$ which form one of the configurations shown in Fig. 6.19.*

The linear weight $W_1(p)$ is defined with respect to the single f-crossing hm (normally, we should denote it by $W_1(p, hm)$). Note that we do not multiply by $w(hm)$. For the positive tetrahedron equation, this wouldn't make any difference. But we will see later that this is forced by the cube equations.

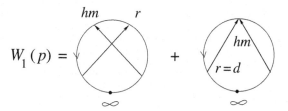

$$W_1(p) =$$

Figure 6.19:　The linear weight W_1.

Let γ be an oriented generic arc in M_T^{reg} which intersects $\Sigma^{(1)}$ only in positive triple crossings.

Definition 6.13 *The evaluation of the 1-cochain $R^{(1)}$ on γ is defined by*

$$R^{(1)}(\gamma) = \sum_{p \in \gamma} \mathrm{sign}(p) W_1(p) \sigma_2 \sigma_1(p)$$

$$+ \sum_{p \in \gamma} \mathrm{sign}(p) z W_2(p)(\sigma_2(p) - \sigma_1(p))$$

where the first sum is over all triple crossings of the global types shown in Fig. 6.15. The second sum is over all triple crossings p which have a distinguished crossing d of type 0.

The following lemma is a "quadratic refinement" of Lemma 6.2.

Lemma 6.4

(1) $W_2(P_1) = W_2(\bar{P}_1) = W_2(P_4) = W_2(\bar{P}_4)$.
(2) $W_2(P_2) = W_2(\bar{P}_2)$.
(3) *Let P_i for some $i \in \{1, 2\}$ be of one of the types shown in Fig. 6.15. Then, $W_1(P_i) = W_1(\bar{P}_i)$.*
(4) *If the f-crossings in P_3 and $\bar{P}_3$ coincide, then $W_2(P_3) = W_2(\bar{P}_3)$.*
(5) *Let f be the new f-crossing in $\bar{P}_3$ with respect to P_3 and let $f = hm$ be the corresponding crossing in P_1. Then, $W_2(\bar{P}_3) - W_2(P_3) = W_2(\bar{P}_3, f)$ and $W_2(\bar{P}_3, f) = W_1(P_1) = W_1(\bar{P}_1)$.*
(6) *Let P_i for some $i \in \{3, 4\}$ be of one of the types shown in Fig. 6.15. Then, either simultaneously $W_1(P_3) = W_1(\bar{P}_3)$ and $W_1(P_4) = W_1(\bar{P}_4)$ or simultaneously $W_1(P_3) = W_1(\bar{P}_3) + 1$ and $W_1(\bar{P}_4) = W_1(P_4) + 1$.*

Proof. The proof is by inspection of all f-crossings, all r-crossings and all hm-crossings in all 24 cases in the figures. We give it in detail in the following. (Those strata which can never contribute are dropped.) It shows in particular that it is necessary to add the degenerate configurations in Figs. 6.16 and 6.19.

I_1: Nothing at all.

I_2: $W_2(P_1) = W_2(\bar{P}_1) = W_2(P_4) = W_2(\bar{P}_4) = 0$. $W_2(P_2) = W_2(\bar{P}_2) = 0$.

I_3: $W_2(P_1) = W_2(\bar{P}_1) = W_2(P_4) = W_2(\bar{P}_4) = 2$. $W_2(P_2) = W_2(\bar{P}_2) = 2$. $W_1(P_1) = W_1(\bar{P}_1) = 2$. $W_1(P_2) = W_1(\bar{P}_2) = 2$. $W_2(P_3) = 0$ and $W_2(\bar{P}_3) = W_2(\bar{P}_3, f) = 2$.

I_4: $W_2(P_1) = W_2(\bar{P}_1) = W_2(P_4) = W_2(\bar{P}_4) = 4$. $W_1(P_1) = W_1(\bar{P}_1) = 1$. $W_1(P_3) = W_1(\bar{P}_3) = 1$. $W_1(P_4) = W_1(\bar{P}_4) = 2$. $W_2(\bar{P}_3) - W_2(P_3) = W_2(\bar{P}_3, f) = 1$.

II_1: $W_1(P_2) = W_1(\bar{P}_2) = 0$.

II_2: $W_2(P_1) = W_2(\bar{P}_1) = W_2(P_4) = W_2(\bar{P}_4) = 2$. $W_1(P_3) = 1$ and $W_1(\bar{P}_3) = 0$. $W_1(P_4) = 1$ and $W_1(\bar{P}_4) = 2$.

II_3: $W_2(P_1) = W_2(\bar{P}_1) = W_2(P_4) = W_2(\bar{P}_4) = 0$. $W_2(P_2) = W_2(\bar{P}_2) = 0$.

II_4: $W_2(P_1) = W_2(\bar{P}_1) = W_2(P_4) = W_2(\bar{P}_4) = 3$. $W_1(P_1) = W_1(\bar{P}_1) = 2$. $W_1(P_2) = W_1(\bar{P}_2) = 2$. $W_2(\bar{P}_3) - W_2(P_3) = W_2(\bar{P}_3, f) = 2$. $W_1(P_4) = W_1(\bar{P}_4) = 1$.

III_1: $W_1(P_3) = W_1(\bar{P}_3) = W_1(P_4) = W_1(\bar{P}_4) = 0$.

III_2: $W_2(P_1) = W_2(\bar{P}_1) = W_2(P_4) = W_2(\bar{P}_4) = 0$. $W_2(P_2) = W_2(\bar{P}_2) = 0$. $W_1(P_3) = W_1(\bar{P}_3) = 0$.

III_3: $W_2(P_2) = W_2(\bar{P}_2) = 3$. $W_1(P_2) = W_1(\bar{P}_2) = 1$.

III_4: $W_2(P_1) = W_2(\bar{P}_1) = W_2(P_4) = W_2(\bar{P}_4) = 2$. $W_1(P_1) = W_1(\bar{P}_1) = 2$. $W_2(P_2) = W_2(\bar{P}_2) = 2$. $W_1(P_2) = W_1(\bar{P}_2) = 2$. $W_2(\bar{P}_3) - W_2(P_3) = W_2(\bar{P}_3, f) = 2$.

IV_1: $W_1(P_1) = W_1(\bar{P}_1) = 0$. $W_2(\bar{P}_3) - W_2(P_3) = W_2(\bar{P}_3, f) = 0$. $W_1(P_3) = W_1(\bar{P}_3) = 1$. $W_1(P_4) = W_1(\bar{P}_4) = 0$.

IV_2: $W_2(P_2) = W_2(\bar{P}_2) = 3$.

IV_3: $W_1(P_1) = W_1(\bar{P}_1) = 1.$ $W_1(P_2) = W_1(\bar{P}_2) = 1.$ $W_2(P_2) = W_2(\bar{P}_2) = 2.$ $W_2(\bar{P}_3) - W_2(P_3) = W_2(\bar{P}_3, f) = 1.$

IV_4: $W_2(P_1) = W_2(\bar{P}_1) = W_2(P_4) = W_2(\bar{P}_4) = 0.$ $W_2(P_2) = W_2(\bar{P}_2) = 0.$ $W_1(P_3) = W_1(\bar{P}_3) = 1.$ $W_2(P_3) = W_2(\bar{P}_3) = 1.$

V_1: $W_1(P_1) = W_1(\bar{P}_1) = 0.$ $W_1(P_2) = W_1(\bar{P}_2) = 0.$ $W_2(\bar{P}_3) - W_2(P_3) = W_2(\bar{P}_3, f) = 0.$

V_2: Nothing at all.

V_3: $W_2(P_1) = W_2(\bar{P}_1) = W_2(P_4) = W_2(\bar{P}_4) = 0.$ $W_2(P_2) = W_2(\bar{P}_2) = 0.$ $W_2(P_3) = W_2(\bar{P}_3) = 0.$

V_4: $W_2(P_1) = W_2(\bar{P}_1) = W_2(P_4) = W_2(\bar{P}_4) = 4.$ $W_2(P_3) = W_2(\bar{P}_3) = 4.$ $W_2(P_4) = W_2(\bar{P}_4) = 4.$ $W_1(P_3) = 3.$ $W_1(\bar{P}_3) = 2.$ $W_1(P_4) = 1.$ $W_1(\bar{P}_4) = 2.$

VI_1: $W_1(P_1) = W_1(\bar{P}_1) = 0.$ $W_1(P_2) = W_1(\bar{P}_2) = 0.$ $W_1(P_4) = W_1(\bar{P}_4) = 0.$ $W_2(\bar{P}_3) - W_2(P_3) = W_2(\bar{P}_3, f) = 0.$

VI_2: $W_1(P_3) = 1.$ $W_1(\bar{P}_3) = 0.$ $W_1(P_4) = 0.$ $W_1(\bar{P}_4) = 1.$

VI_3: $W_2(P_2) = W_2(\bar{P}_2) = 4.$

VI_4: $W_2(P_1) = W_2(\bar{P}_1) = W_2(P_4) = W_2(\bar{P}_4) = 0.$ $W_2(P_2) = W_2(\bar{P}_2) = 0.$ $W_2(P_3) = W_2(\bar{P}_3) = 0.$ □

Proposition 6.3 *Let m be the meridian for a positive quadruple crossing. Then, $R^{(1)}(m) = 0$.*

Proof. Points (1) to (5) in Lemma 6.4 imply that we can repeat word for word (without the grading) the proof of Proposition 6.2 but with the coefficient 1 of $\sigma_2\sigma_1(p)$ replaced by $W_1(p)$ and the coefficient $W(p)$ of $z(\sigma_2(p) - \sigma_1(p))$ replaced by $W_2(p)$.

The only new feature is point (6) in Lemma 6.4. If $W_1(P_3) = W_1(\bar{P}_3) + 1$, then $P_3 - \bar{P}_3$ contribute to $R^{(1)}$ the element associated with the tangle on the left-hand side in Fig. 6.20 and $P_4 - \bar{P}_4$ contribute the element associated with the tangle on the right-hand side in Fig. 6.20. But these contributions cancel out (even without using the skein relations). □

Figure 6.20: Cancellation in (6) of Lemma 6.4.

We have proven that $R^{(1)}$ is a solution of the positive global tetrahedron equation.

Remark 6.5 *For "dual" 1-cocycles, we have of course to construct the weights W_1 and W_2 from the second formula for v_2 of Polyak and Viro (compare Fig. 1.3) by using the "heads" of the arrows.*

6.6 Solution with Linear Weight of the Cube Equations

We start with some observations and notations. The identity $\sigma_2(p) - \sigma_1(p) = \sigma_2^{-1}(p) - \sigma_1^{-1}(p)$ is sometimes useful to simplify calculations. We have given names to the global types of strata of $\Sigma_{\mathrm{tri}}^{(1)}$ in Fig. 6.6. *The union of all strata of the same type is denoted simply by this name.* The weights and the partial smoothings in the case of positive triple crossings were determined in the last two sections. We have to consider now all other local types of triple crossings together with the self-tangencies.

Definition 6.14 *The weight $W(p)$ is the same as in Definition 6.8 for all local types of triple crossings p.*

For a triple crossing p of a given local type from $\{1, \ldots, 8\}$, we denote by $T_{r_a}(type)(p) = T_{l_c}(type)(p)$ and $T(type)(p)$ the corresponding partial smoothing of p in the skein module. We will determine these partial smoothings from the cube equations.

Definition 6.15 *We distinguish four types of self-tangencies, denoted by $II_0^+, II_0^-, II_1^+, II_1^-$. Here, "0" or "1" is the type of the distinguished crossing d and "+" stands for opposite tangent directions of the two branches and "−" stands for the same tangent direction of the two branches.*

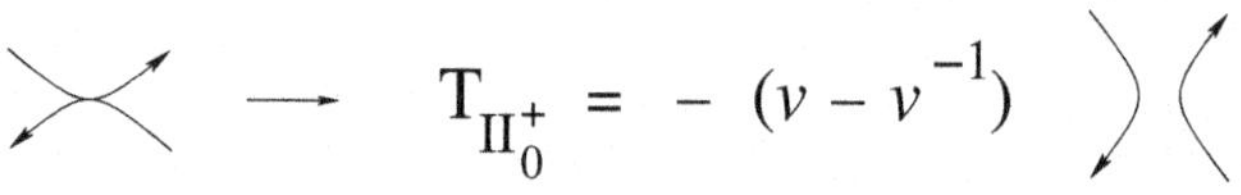

$$\asymp \quad \longrightarrow \quad T_{II_0^+} = -(v - v^{-1}) \quad \big)\big($$

Figure 6.21:   Partial smoothing of a self-tangency II_0^+.

Definition 6.16 *The partial smoothing $T_{II_0^+}(p)$ of a self-tangency with opposite tangent direction is defined in Fig. 6.21. The partial smoothing $T_{II_0^-}(p)$ of a self-tangency with equal tangent direction is defined as $-zP_T$ (where P_T is the element represented by the original tangle T in the HOMFLYPT skein module $S(\partial T)$).*

Note that the distinguished crossing d has to be of type 0 in both cases.

Definition 6.17 *Let p be a self-tangency. Its contribution to $R_{\mathrm{reg}}^{(1)}$ is defined by*

$$R_{\mathrm{reg}}^{(1)} = \mathrm{sign}(p)W(p)T_{II_0^+}(p), \ \textit{respectively,} \ \mathrm{sign}(p)W(p)T_{II_0^-}(p).$$

Its contribution to $R^{(1)}$ is defined by

$$R^{(1)} = \mathrm{sign}(p)W_2(p)T_{II_0^+}(p), \ \textit{respectively,} \ \mathrm{sign}(p)W_2(p)T_{II_0^-}(p).$$

(Don't forget that in the case of $R^{(1)}$ the HOMFLYPT invariants have to be normalised by $v^{-w(T)}$.)

There are exactly two local types of self-crossings with opposite tangent direction as well as for equal tangent direction. The corresponding strata from $\Sigma_{\mathrm{tan}}^{(1)}$ are adjacent in $\Sigma_{\mathrm{self-flex}}^{(2)}$.

Lemma 6.5 *Let m be the meridian of $\Sigma_{\mathrm{self-flex}}^{(2)}$ (compare Section 6.2). Then, $R_{\mathrm{reg}}^{(1)}(m) = 0$ and $R^{(1)}(m) = 0$.*

Proof.   We consider the case with opposite tangent direction. We show the unfolding (i.e., the intersection of a meridional disc with Σ in M_T) in Fig. 6.22 (compare Ref. [19]). There are two cases: either all three crossings in the picture are of type 1 or all three are of type 0. In the first case, both self-tangencies in the unfold-

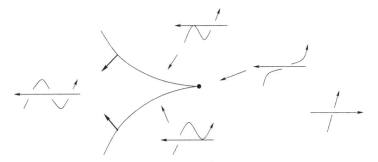

Figure 6.22: The meridional disc for a self-tangency in a flex.

$$- W_2\left(v - v^{-1}\right)\left(\text{—} \right) + W_2\left(v - v^{-1}\right)\left(\text{—}\right) = 0$$

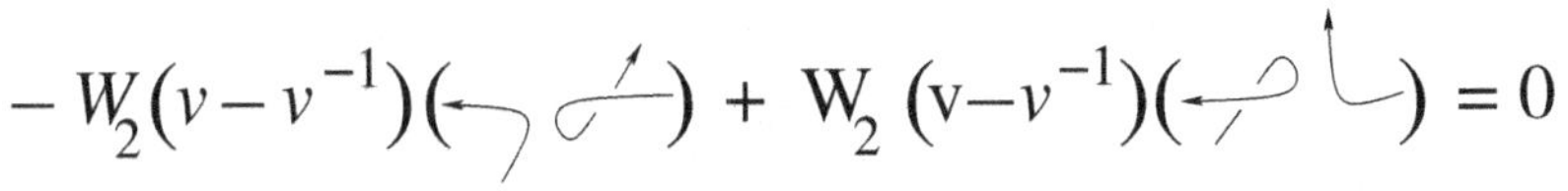

Figure 6.23: $R^{(1)}$ vanishes on the meridian of a self-tangency in a flex.

ing do not contribute to $R^{(1)}_{\text{reg}}$ and to $R^{(1)}$ (because d has to be of type 0). In the second case, both self-tangencies have different sign but the same weight $W(p)$ because the crossing which is not in d is not an f-crossing. But they also have the same weight $W_2(p)$. Indeed, the crossing which is not in d has for both self-crossings the same writhe and if it is an r-crossing for an f-crossing for one of the self-tangencies, then this is the case for the other self-tangency too. It follows that $R^{(1)}_{\text{reg}}(m) = 0$ and $R^{(1)}(m) = 0$ as shown in Fig. 6.23. The case with equal tangent direction is analogue (the third crossing can never be an f-crossing).

□

This lemma implies that we can restrict ourself to consider in the cube equations only one of the two local types of self-tangencies. The local types of triple crossings are shown in Fig. 2.2. The diagrams which correspond to the edges of the graph Γ (compare Section 6.2) are shown in Fig. 6.24. We show the corresponding graph Γ in Fig. 3.4. The unfolding of, e.g., the edges 1–5 is shown in Fig. 6.24 (compare Ref. [19]).

Observation 6.1 *The diagrams corresponding to the two vertices's of an edge differ just by the two crossings of the self-tangency which replace each other in the triangle as shown in Fig. 6.25. Consequently,*

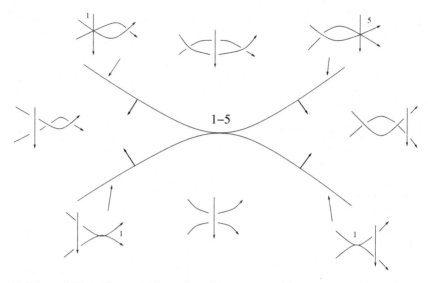

Figure 6.24:   The unfolding of a self-tangency with a transverse branch.

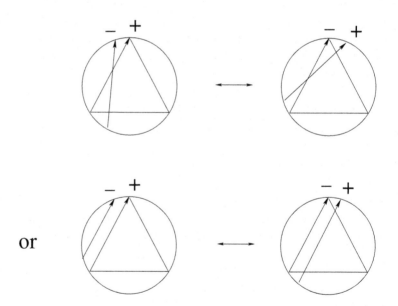

Figure 6.25:   The two vertices of an edge of Γ.

the two vertices of an edge always have the same global type and always different local types and the two crossings which are interchanged are simultaneously f-crossings and have the same grading A. Moreover, the two self-tangencies in the unfolding (of an edge) can contribute non-trivially to $R_{\text{reg}}^{(1)}$ if they have different weights $W(p)$ or if their partial smoothings are not isotopic. The latter happens exactly for the edges "1–6" and "2–8" (where the third branch passes between the two branches with opposite orientation of the self-tangency).

It follows from Observation 6.1 and Lemma 6.5 that we have to solve the cube equations for the graph Γ exactly six times: one solution for each global type of triple crossings.

Definition 6.18 *The partial smoothings for the local and global types of triple crossings are given in Figs. 6.26 and 6.27 (remember that* mid *denotes the ingoing middle branch for a star-like triple crossing).*

Definition 6.19 *Let s be a generic oriented arc in M_T^{reg} (with a fixed abstract closure $T \cup \sigma$ to an oriented circle). Let $A \subset \partial T$ be a given*

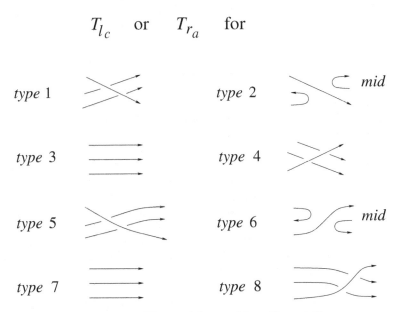

Figure 6.26: The partial smoothings T_{l_c} and T_{r_a}.

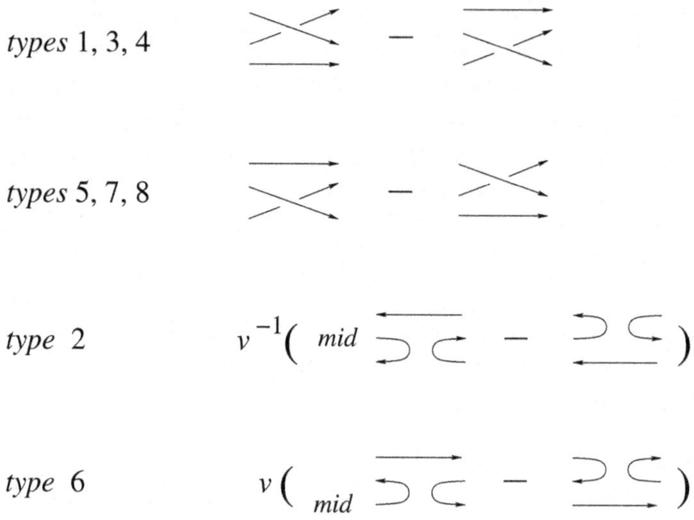

types 1, 3, 4

types 5, 7, 8

type 2 $v^{-1}\big(\ mid\ \ \ -\ \ \ \big)$

type 6 $v\big(\ _{mid}\ \ \ -\ \ \ \big)$

Figure 6.27: The partial smoothings T (type).

grading. The 1-cochain $R_{\mathrm{reg}}^{(1)}(A)$ *is defined by*

$$R_{\mathrm{reg}}^{(1)}(A)(s) = \sum_{p \in s \cap l_c} \mathrm{sign}(p) T_{l_c}(\mathrm{type})(p)$$

$$+ \sum_{p \in s \cap r_a} \mathrm{sign}(p) T_{r_a}(\mathrm{type})(p)$$

$$+ \sum_{p \in s \cap r_a} \mathrm{sign}(p) z W(p) T(\mathrm{type})(p)$$

$$+ \sum_{p \in s \cap r_b} \mathrm{sign}(p) z W(p) T(\mathrm{type})(p)$$

$$+ \sum_{p \in s \cap l_b} \mathrm{sign}(p) z W(p) T(\mathrm{type})(p)$$

$$+ \sum_{p \in s \cap II_0^+} \mathrm{sign}(p) W(p) T_{II_0^+}(p)$$

$$+ \sum_{p \in s \cap II_0^-} \mathrm{sign}(p) W(p) T_{II_0^-}(p).$$

Here, all weights $W(p)$ are defined only over the f-crossings f with $\partial f = A$ and in the first two sums (i.e., for T_{l_c} and T_{r_a}), we require that $\partial(hm) = A$ for the triple crossings.

Note that the stratum r_a contributes in two different ways.

This turns out to be the solution of the cube equations with linear weight.

Proposition 6.4 *Let m be a meridian of $\Sigma^{(2)}_{\text{trans}-\text{self}}$ or a loop in Γ and let A be a given grading. Then, $R^{(1)}_{\text{reg}}(A)(m) = 0$ for the partial smoothings given in Definitions* 6.17 *and* 6.18.

The proof will be in Figs. 6.28–6.65.

Proof. First of all, we observe that the two vertices of an edge share the same weight $W(p)$. The weight could only change if the foot of an f-crossing slides over the head of the crossing d. But this is not the case as it is shown in Fig. 6.25. Indeed, the foot of the crossing which changes in the triangle cannot coincide with the head of d because the latter coincides always with the head of another crossing.

$$l{:}1\text{--}5$$

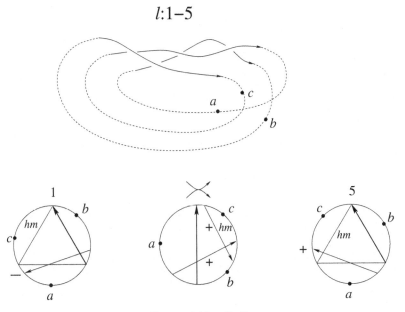

Figure 6.28: l1–5.

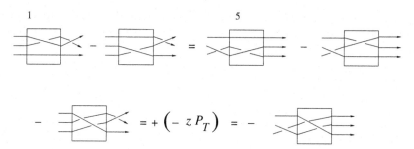

Figure 6.29:   Smoothings for $l1$–5.

l: 7–2

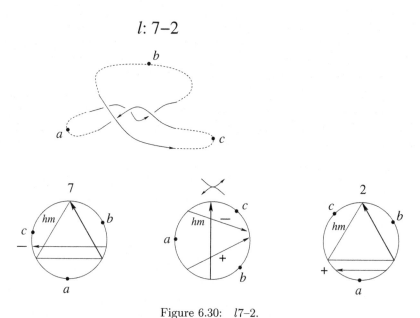

Figure 6.30: $l7$–2.

It suffices of course to consider only the four crossings involved in an edge because their positions with respect to all other crossings do not change.

Our strategy is the following: we try to solve the equations separately for the partial smoothings with constant weight of triple crossings and the partial smoothings with linear weight. However, they cannot be separated completely. It turns out that the two self-tangencies in the unfolding for the edge "1–6" as well as for the edge

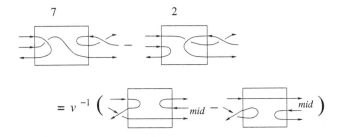

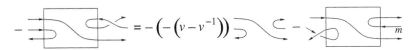

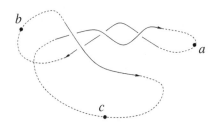

Figure 6.31:   Smoothings for $l7$–2.

l: 3–6

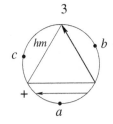

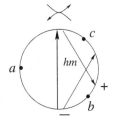

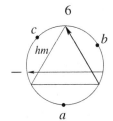

Figure 6.32: $l3$–6.

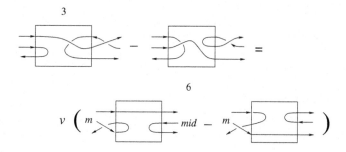

Figure 6.33:   Smoothings for $l3$–6.

l: 1–7

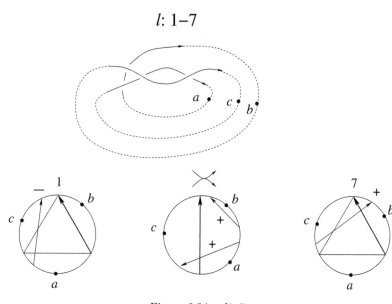

Figure 6.34: $l1$–7.

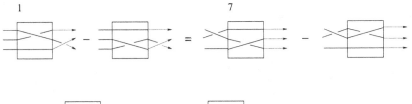

Figure 6.35:   Smoothings for $l1$–7.

$l: 5$–3

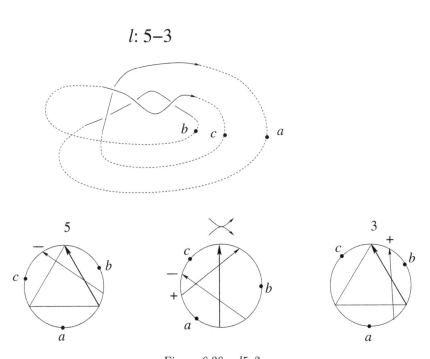

Figure 6.36: $l5$–3.

"2–8" always have the same weight $W(p)$. Therefore, they will contribute to the smoothings with linear weight of triple crossings. For the edges "2–7" and "3–6," the smoothings of the two self-tangencies are isotopic. However, the weight $W(p)$ changes by 1. Consequently, these self-tangencies have to contribute to the partial smoothings with constant weight of triple crossings. For the edges "2–5" and

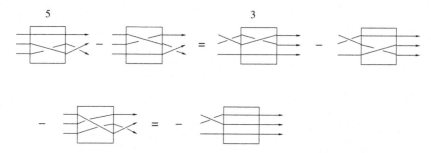

Figure 6.37: Smoothings for *l*5–3.

l: 7–4

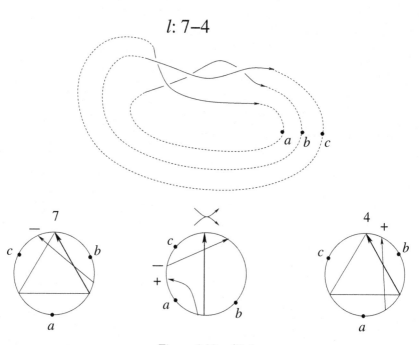

Figure 6.38: *l*7–4.

"4–6," the two self-tangencies have the same weight $W(p)$ and do not contribute.

We start with the solutions for triple crossings of global type l. In the figures, we show the Gauss diagrams of the two triple crossings together with the points at infinity and the Gauss diagram of just one of the two self-tangencies. The Gauss diagram of the second self-tangency is derived from the first one in the following way: the

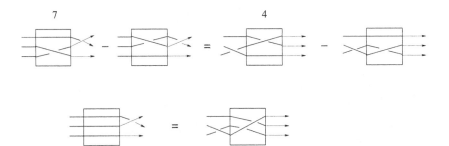

Figure 6.39: Smoothings for $l7$–4.

l: 4–6

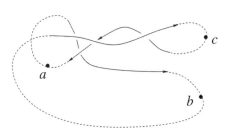

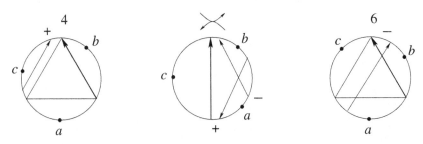

Figure 6.40: $l4$–6.

two arrows slide over the arrow d but their mutual positions do not change. We show an example in Fig. 6.65. Note that in the thick part of the circle there aren't any other heads or foots of arrows.

To each figure of an edge corresponds a second figure. In the first line on the left-hand side, we have the smoothing with constant weight which is already known. (Of course, we draw only the case

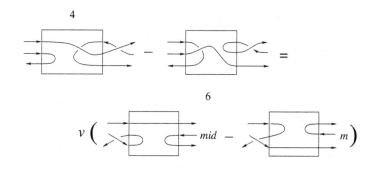

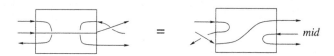

Figure 6.41:   Smoothings for $l4$–6.

l: 3–8

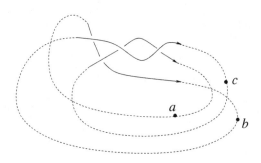

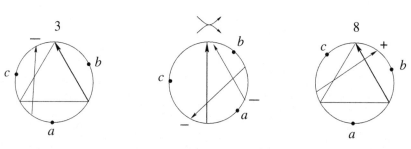

Figure 6.42: $l3$–8.

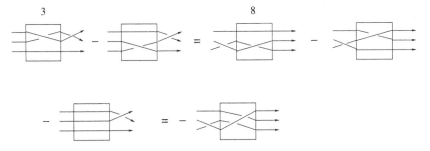

Figure 6.43: Smoothings for $l3$–8.

when the triple crossing contributes, i.e., global type l_b or l_c.) This determines the new smoothing on the right-hand side. We put the smoothings in rectangles for better visualising them. In the second line, we give the corresponding smoothings with linear weight (but we drop the common factor $W(p)$). Here, we often use the identity mentioned at the beginning of this section.

After establishing the partial smoothings for all eight local types of triple crossings, we have to verify that the solution is consistent for the loops in Γ. We do this for the boundary of the three different types of 2-faces "1–7–2–5," "1–6–4–7" and "1–5–3–6" of the cube. The rest of the 2-faces are completely analogous, and we left the verification to the reader.

We proceed then in exactly the same way for the global type r. But note the difference which comes from the fact that we have broken the symmetry: the global type r_a contributes both with constant weight as well as with linear weight.

The rest of the proof is in Figs. 6.28–6.65. ☐

Remark 6.6 *It is not hard to see that one can find a solution* $R^{(1)}_{\text{reg}}(A)(m) = 0$ *for the meridians of* $\Sigma^{(2)}_{\text{trans}-\text{self}}$ *without using self-tangencies at all. But this solution does not survive for the loops in* Γ. *Hence, the contributions of the self-tangencies with opposite tangent direction are necessary.*

However, our solution of the cube equations is not unique!

The reader can easily verify (using our figures) that the following is another solution of the cube equations: we do not use self-tangencies of type II_0^- *at all. But for the contributions with constant weight of the triple crossings in the boundary of the* 2-*face* "2–5–3–8,"

l: 4–8

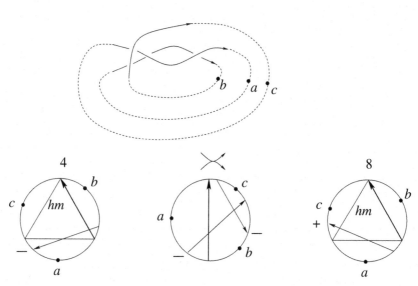

Figure 6.44: l4–8.

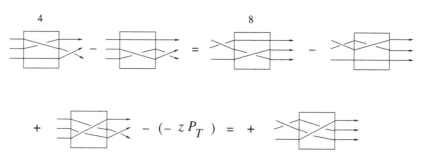

Figure 6.45:   Smoothings for l4–8.

we add two additional negative crossings between two branches as shown in Fig. 6.66. Moreover, we add $\operatorname{sign}(p)W(p)zP_T$ to the contributions of the self-tangencies of type II_0^+.

It turns out that the new 1-cocycle also has the scan-property. We don't know whether it contains the same information as $R_{\text{reg}}^{(1)}(A)$. However, we will need the simpler partial smoothings given above in the construction of the 1-cocycle $\bar{R}^{(1)}$ (compare Lemma 6.10).

l: 5–2

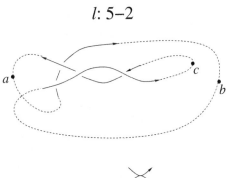

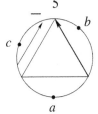 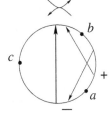 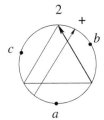

Figure 6.46: *l*5–2.

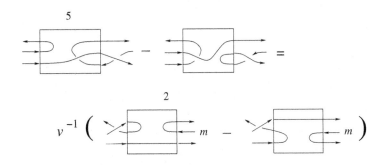

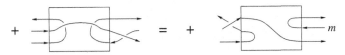

Figure 6.47: Smoothings for *l*5–2.

l: 1–6

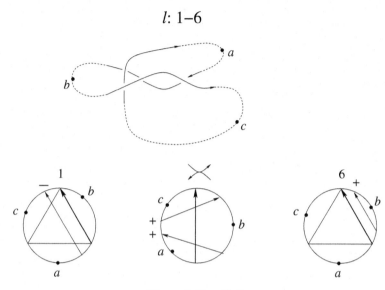

Figure 6.48: $l1$–6.

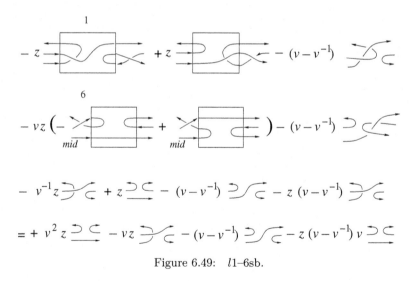

Figure 6.49: $l1$–6sb.

Figure 6.50: $l1$–6sc.

l: 8–2

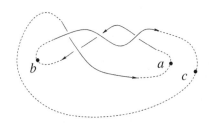

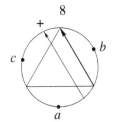 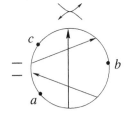 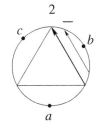

Figure 6.51: *l*8–2.

r: 1–7

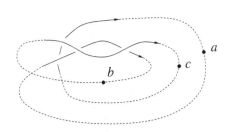

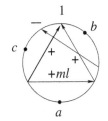 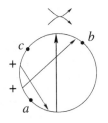 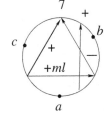

Figure 6.52: *r*1–7.

r: 1–5

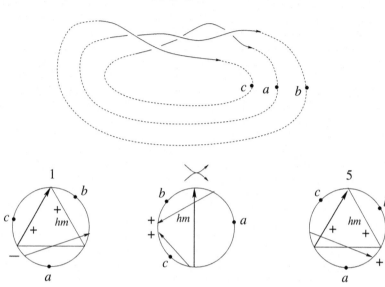

Figure 6.53:   r1–5.

r: 5–3

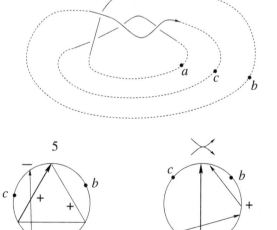
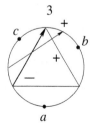

Figure 6.54:   r5–3.

r: 4–8

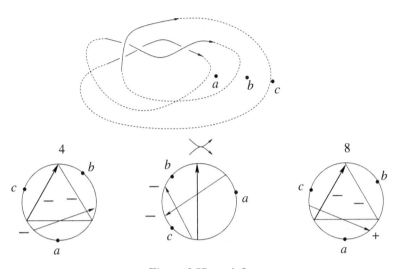

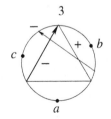

Figure 6.55: r4–8.

r: 3–8

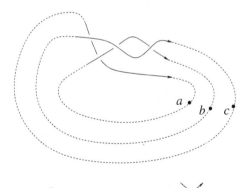

Figure 6.56: r3–8.

r: 7–2

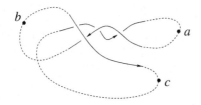

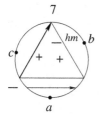 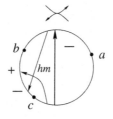 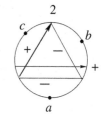

Figure 6.57:   r7–2.

r: 3–6

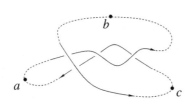

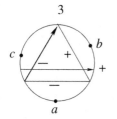 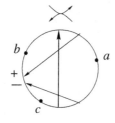 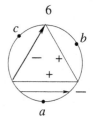

Figure 6.58:   r3–6.

r: 8–2

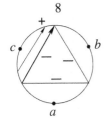
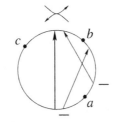
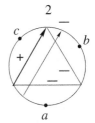

Figure 6.59:   *r*8–2.

r: 1–6

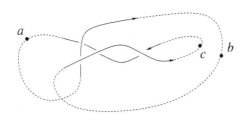

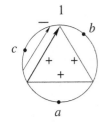
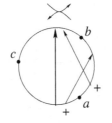
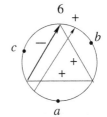

Figure 6.60:   *r*1–6.

r: 1–6a

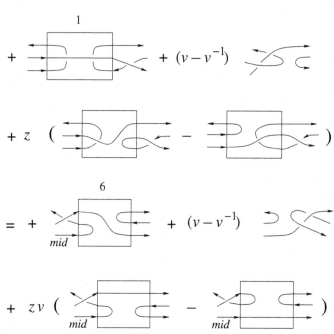

Figure 6.61:   Smoothings for r1–6.

r: 5–2

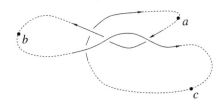

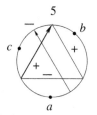 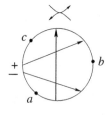 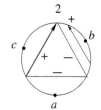

Figure 6.62:   r5–2.

r: 4–6

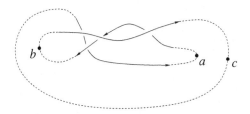

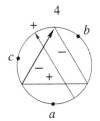

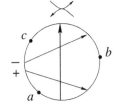

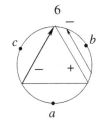

Figure 6.63:   r4–6.

r: 7–4

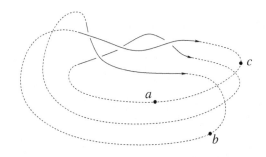

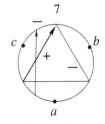

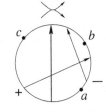

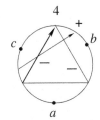

Figure 6.64:   r7–4.

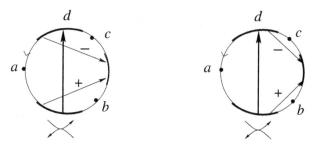

Figure 6.65:   The two self-tangencies for the edge $l7$–2.

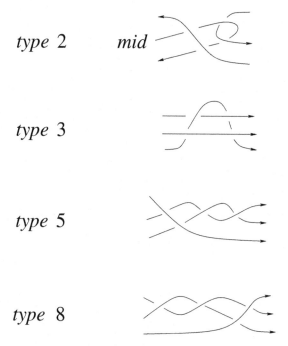

Figure 6.66:   Alternative solution for the cube equations without type II_0^-.

6.7   Solution with Quadratic Weight of the Cube Equations

We have to replace the constant weight 1 by the linear weight $W_1(p)$ and the linear weight $W(p)$ by the quadratic weight $W_2(p)$. All partial smoothings are exactly the same as in the previous section besides an adjustment of the weights by a constant which depends on the global and on the local type.

The weights $W_1(p)$ and $W_2(p)$ for positive triple crossings (local type 1) were introduced in Definitions 6.12 and 6.11, respectively.

Definition 6.20 *For the global types l_b, l_c and r_b, the weights $W_1(p)$ and $W_2(p)$ do not depend on the local type. For the global type r_a, the weights are defined as follows:*

$$W_1(\text{type1}) = W_1(\text{type3}) = W_1(\text{type7}) = W_1(\text{type8}) = W_1(p)$$
$$W_2(\text{type1}) = W_2(\text{type3}) = W_2(\text{type7}) = W_2(\text{type8}) = W_2(p)$$
$$W_1(\text{type2}) = W_1(\text{type5}) = W_1(p) - 1$$
$$W_1(\text{type4}) = W_1(\text{type6}) = W_1(p) + 1$$
$$W_2(\text{type4}) = W_2(\text{type5}) = W_2(p) - 1$$
$$W_2(\text{type2}) = W_2(\text{type6}) = W_2(p) + 1.$$

Definition 6.21 *Let s be a generic oriented arc in M_T^{reg} (with a fixed abstract closure $T \cup \sigma$ to an oriented circle). The 1-cochain $R^{(1)}$ is defined by*

$$R^{(1)}(s) = \sum_{p \in s \cap l_c} \text{sign}(p) W_1(p) T_{l_c}(\text{type})(p)$$

$$+ \sum_{p \in s \cap r_a} \text{sign}(p) [W_1(\text{type})(p) T_{r_a}(\text{type})(p)$$

$$+ z W_2(\text{type})(p) T(\text{type})(p)]$$

$$+ \sum_{p \in s \cap r_b} \text{sign}(p) z W_2(p) T(\text{type})(p)$$

$$+ \sum_{p \in s \cap l_b} \text{sign}(p) z W_2(p) T(\text{type})(p)$$

$$+ \sum_{p \in s \cap II_0^+} \text{sign}(p) W_2(p) T_{II_0^+}(p)$$

$$+ \sum_{p \in s \cap II_0^-} \text{sign}(p) W_2(p) T_{II_0^-}(p).$$

Here, all partial smoothings are exactly the same as in the previous section.

Proposition 6.5 *Let m be a meridian of $\Sigma^{(2)}_{trans-self}$ or a loop in Γ. Then, $R^{(1)}(m) = 0$.*

Proof.   For the global type l, the proof is exactly the same as in the case of linear weight. Indeed, the fourth arrow which is not in the triangle is always almost identical with an arrow of the triangle. Consequently, in the case l_c, it cannot be an r-crossing with respect to hm. In the case l_b, it could be an r-crossing with respect to some f-crossing. But the almost identical arrow in the triangle would be an r-crossing for the same f-crossing too. Their contributions cancel out because they have different writhe.

The mutual position of the two arrows for a self-tangency does not change. Only the position with respect to the distinguished crossing d changes. But this does not count because d consists in fact of two crossings with opposite writhe. It remains to consider the edges "2–7" and "3–6" for l_c, where one of the two self-tangencies has a new f-crossing with respect to the other self-tangency. But one sees immediately from the figures that this new f-crossing is exactly the crossing hm in the triple crossings. Consequently, all three terms in the first line of the figures are just multiplied by the same $W_1(p)$ and the equations are still satisfied.

Exactly the same arguments apply in the case r_b too. In the case r_c, there are no contributions at all. The differences appear in the case r_a.

We examine the figures (don't forget the degenerate configurations):

edge "1–7": type 1 and type 7 share the same W_1 and W_2

edge "3–8": type 3 and type 8 share the same W_1 and W_2

edge "4–6": same W_1 but $W_2 + 1$ for type 4 and $W_2 - 1$ for type 6

edge "5–2": same W_1 but $W_2 + 1$ for type 5 and $W_2 - 1$ for type 2

edge "1–5": $W_1 - 1$ and $W_2 - 1$ for type 1 with respect to type 5

edge "7–4": $W_1 - 1$ and $W_2 + 1$ for type 4 with respect to type 7

edge "5–3": $W_1 + 1$ and $W_2 + 1$ for type 5 with respect to type 3

edge "4–8": $W_1 - 1$ and $W_2 + 1$ for type 4 with respect to type 8

edge "7–2": $W_1 + 1$ and $W_2 - 1$ for type 2 with respect to type 7

edge "3–6": $W_1 - 1$ and $W_2 - 1$ for type 6 with respect to type 3

edge "8–2": $W_1 - 1$ and $W_2 + 1$ for type 8 with respect to type 2
edge "1–6": $W_1 + 1$ and $W_2 + 1$ for type 1 with respect to type 6.

For both self-tangencies in "8–2" as well as in "1–6," we have the same W_2 as in type 8, respectively, type 1.

It follows now easily that all cube equations are satisfied with the adjustments of the weights given in Definition 6.20. □

6.8 The 1-Cocycle $R_{\mathrm{reg}}^{(1)}$, the Scan-Property and First Examples

The 1-cochain $R_{\mathrm{reg}}^{(1)}(A)$ was defined in Definition 6.19 and the 1-cochain $R^{(1)}$ in Definition 6.21. We have proven that they satisfy the global positive tetrahedron equation as well as the cube equations. Moreover, we have proven that their value is 0 on the meridians of the strata of codimension 2 which correspond to a self-tangency in an ordinary flex as well as to the transverse intersections of two strata of codimension 1. These are the first four points from the list in Section 6.2. These are the only strata of codimension 2 which appear in generic homotopies of generic regular isotopies in M_T^{reg} (compare Ref. [19]). Consequently, we have proven the following theorem.

Theorem 6.2 *The 1-cochains $R_{\mathrm{reg}}^{(1)}(A)$ and $R^{(1)}$ are 1-cocycles in M_T^{reg}.*

It is difficult to apply this theorem. The natural loops *rot* and *fh* in M_T are not in M_T^{reg}. Of course, we could approximate them by loops in M_T^{reg} using Whitney tricks. However, this approximation is not unique and leads to very long calculations. But we can apply it to the loop which consists of dragging a knot through itself by a regular isotopy (compare the Introduction). We have calculated by hand that $R_{\mathrm{reg}}^{(1)}(A)(\mathrm{drag}3_1^+) = 0$. So, we don't know whether or not $R_{\mathrm{reg}}^{(1)}(A)$ represents the trivial cohomology class in M_T^{reg}.

Remark 6.7 *Even the specialisation of $R_{\mathrm{reg}}^{(1)}(A)$ for $v = 1$ cannot be extended to a 1-cocycle for isotopies instead of regular isotopies.*

Indeed, the intersection with a stratum $\Sigma_{\mathrm{cusp}}^{(1)}$ does not change the calculation of the Conway invariants for the intersections with the other strata of $\Sigma^{(1)}$. There are two types of strata in $\Sigma^{(1)}_{\text{cusp}}$: the new crossing could be of type 0 or of type 1. Let's call them $\Sigma^{(1)}_{\text{cusp}\,0}$ and $\Sigma^{(1)}_{\text{cusp}\,1}$, respectively. The value of the 1-cocycle on a meridian of $\Sigma^{(1)} \cap \Sigma^{(1)}_{\text{cusp}\,0}$ is 0 because we have set $v = 1$. However, it is not controllable at all on a meridian of $\Sigma^{(1)} \cap \Sigma^{(1)}_{\text{cusp}\,1}$. Indeed, the new crossing of type 1 could be an f-crossing for the stratum in $\Sigma^{(1)}$. Moreover, the 1-cocycle can behave uncontrollable on a meridian of $\Sigma^{(2)}_{\text{trans}-\text{cusp}}$ (compare Section 6.2). We show an example in Fig. 6.67.

The good news are that $R^{(1)}_{\text{reg}}(A)$ and $R^{(1)}$ turn out to have the *scan-property*.

Lemma 6.6 *Let T be a diagram of a string link and let t be a Reidemeister move of type II for T. Then, the contribution of t to $R^{(1)}_{\text{reg}}(A)$, respectively, to $R^{(1)}$ does not change if a branch of T is moved under t from one side of t to the other.*

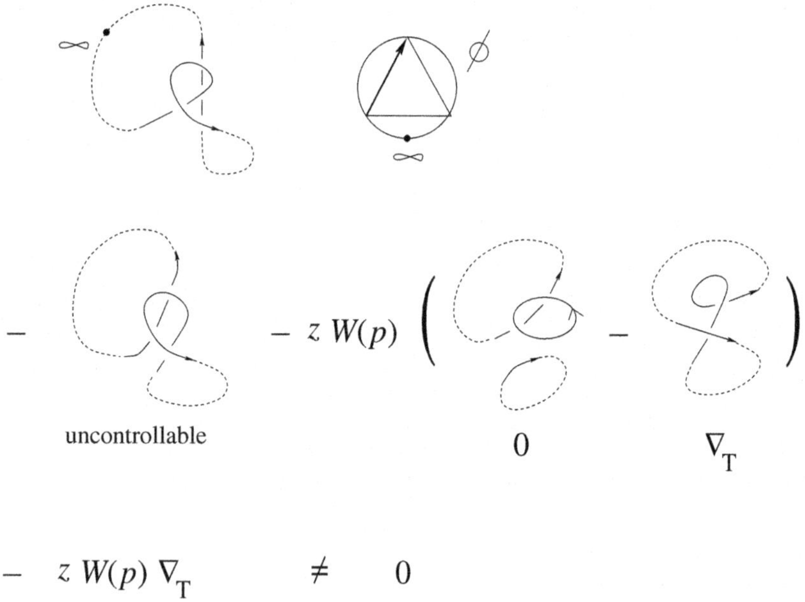

Figure 6.67:   The regular 1-cocycle from the Conway polynomial is not controllable on meridians of $\Sigma^{(2)}_{\text{trans}-\text{cusp}}$.

Proof. Besides d of t, there are two crossings involved with the branch which goes under t. Moving the branch under t from one side to the other slides the heads of the corresponding arrows over d, see, e.g., Figs. 6.34 and 6.52 as well as Figs. 6.40 and 6.63. Consequently, the f-crossings and hence $W(p)$ do not change. Note that the mutual position of the two arrows does not change. Consequently, there are no new r-crossings and $W_2(p)$ does not change. On the other hand, it is clear that the partial smoothings of t on both sides are regularly isotopic. $\qquad\square$

The contribution of a Reidemeister I move will be defined in the next section (Definition 6.22). It does not depend on the place in the diagram and hence it has the scan-property.

We are now ready to prove Theorem 6.2.

Proof of Theorem 6.2. Let T be a diagram of a string link and let s be a regular isotopy which connects T with a diagram T'. We consider the loop-$s \circ \text{-} scan(T') \circ s \circ scan(T)$ in M_T^{reg}. This loop is contractible in M_T^{reg} because s and $scan$ commute and hence $R_{\text{reg}}^{(1)}(A)$ and $R^{(1)}$ vanish on this loop. Consequently, it suffices to prove that each contribution of a Reidemeister move t in s cancels out with the contribution of the same move t in $-s$ (the signs are of course opposite). The difference for the two Reidemeister moves is in a branch which has moved under t. The partial smoothings are always regularly isotopic. Hence, it suffices to study the weights. If t is a positive triple crossing, then the weights are the same just before the branch moves under t and just after it has moved under t. This follows from the fact that for the positive global tetrahedron equation the contribution from the stratum $-P_2$ cancels out with that from the stratum $\bar{P}_2$ (compare Sections 6.4 and 6.5). If we move the branch further away, then the invariance follows from the already proven fact that the values of the 1-cocycles do not change if the loop passes through a stratum of $\Sigma^{(1)} \cap \Sigma^{(1)}$. We use now again the graph Γ. The meridian m which corresponds to an arbitrary edge of Γ is a contractible loop in M_T^{reg} no matter what the position of the branch which moves under everything is. Let's take an edge where one vertex is a triple crossing of type 1. We know from Lemma 6.6 that the contributions of the self-tangencies do not depend on the position of the moving branch. Consequently, the contribution of the other vertex of the edge doesn't change neither because the contributions from all four Reidemeister moves together sum up to 0. Using the fact that the graph Γ is connected, we obtain the invariance with respect to the position of the moving branch for all (regular) Reidemeister moves t. $\qquad\Box$

Note that $R_{\text{reg}}^{(1)}(A)$ and $R^{(1)}$ do not have the scan-property for a branch which moves over everything else because the contributions of the strata P_3 and $-\bar{P}_3$ in the positive tetrahedron equation do not cancel out at all.

We start with very simple examples and we give all details of the calculations in order to make it easier for the reader to become familiar with the 1-cocycles $R_{\text{reg}}^{(1)}(A)$ and $R^{(1)}$. In all examples, we give names $x_1, x_2, \ldots$ to the Reidemeister moves and sometimes we give names $c_1, c_2, \ldots$ to the crossings too.

Example 6.1 Let $T = \sigma_1$. There is a unique closure to a circle and there are two choices for the point at infinity. We chose the small curl on the second branch as shown in Fig. 6.68. The Reidemeister move x_1 is a self-tangency with equal tangent direction. The Reidemeister move x_2 is a positive triple crossing (type 1) with sign$(x_2) = -1$. We see immediately from the Gauss diagrams that $W(x_2) = W_2(x_2) = 0$ in the case ∞_1. Consequently, for the choice ∞_1, we obtain

$$R_{\text{reg}}^{(1)}(\text{scan}_2(\sigma_1)) = R^{(1)}(\text{scan}_2(\sigma_1)) = 0.$$

Let us consider the case ∞_2. The crossing hm in the triangle is the only f-crossing. We obtain $\partial(hm) = \infty_2$, $W(x_2) = 1$, $W_2(x_2) = 0$ (remember that the crossing d here is an r-crossing too) and $W_1(x_2) = 0$ too. $W(x_1) = 1$ with the same grading as hm and $W_2(x_1) = 0$ because d is a positive and a negative crossing. Consequently,

$$R_{\text{reg}}^{(1)}(A = \partial\sigma_1)(\text{scan}_2(\sigma_1)) = R^{(1)}(\text{scan}_2(\sigma_1)) = 0.$$

The calculation of $R_{\text{reg}}^{(1)}(A = \infty_2)(\text{scan}_2(\sigma_1))$ is given in Fig. 6.69 (by using Definitions 6.6, 6.7, 6.14 and 6.15). We obtain

$$R_{\text{reg}}^{(1)}(\infty_2)(\text{scan}_2(\sigma_1)) = -(2v - v^{-1} + vz^2) \cdot 1.$$

Here, 1 and σ_1 are the generators of the skein module $S(\partial\sigma_1)$. It is amazing that the coefficient of the generator 1 is just minus the HOMFLYPT polynomial of the right trefoil.

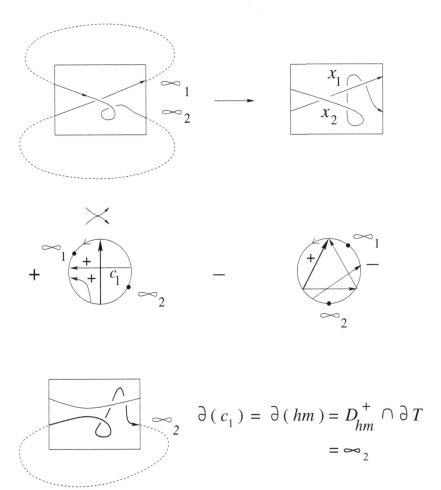

$$\partial\,(\,c_1\,) \;=\; \partial\,(\,hm\,) \;=\; D^{+}_{hm}\cap\partial\,T$$

$$= \infty_2$$

Figure 6.68: The two Reidemeister moves in Example 6.1 for scan_2.

Let us consider the curl on the first branch. We show the corresponding regular isotopy in Fig. 6.70. In the case ∞_1, we obtain $W(x_1) = W_2(x_1) = 0$ and $W(x_2) = W_2(x_2) = 0$. Consequently,

$$R^{(1)}_{\text{reg}}(\text{scan}_1(\sigma_1)) = R^{(1)}(\text{scan}_1(\sigma_1)) = 0.$$

In the case ∞_2, the move x_1 does not contribute because d is of type 1. $\partial(hm) = \infty_2$ for the move x_2 and $W_1(x_2) = 0$. Consequently,

$$R^{(1)}(\text{scan}_1(\sigma_1)) = 0.$$

$$+ \ (-zP_T \ ) \quad - \quad$$

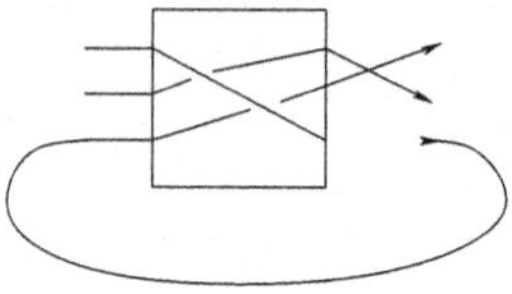

$$- \ z \ \left(\quad - \quad \right)$$

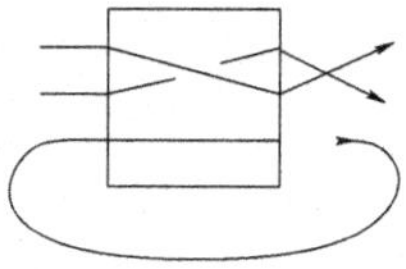

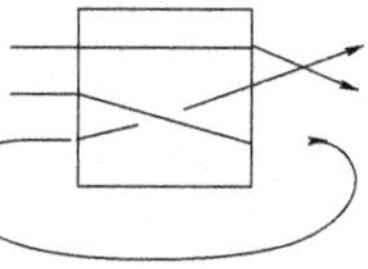

$$= \ -zv\,\sigma_1 \ - \ v\cdot 1 \ - \ z\delta\cdot 1 \ + \ zv\,\sigma_1^{-1}$$

$$= \ (-2v + v^{-1} - v\,z^2 \ )\cdot 1$$

Figure 6.69:   The calculation of $R_{\text{reg}}^{(1)}(\text{scan}_2)$ in Example 6.1.

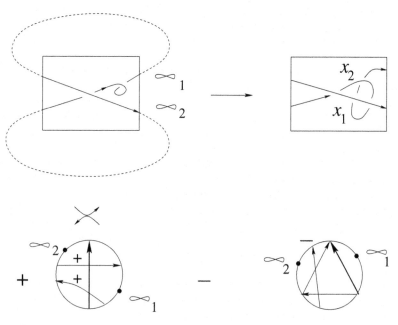

Figure 6.70:   The two Reidemeister moves in Example 6.1 for scan_1.

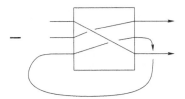

Figure 6.71: Calculation of $R^{(1)}_{\text{reg}}(\infty_2)(\text{scan}_1)$.

The partial smoothing of the triple crossing x_2 is shown in Fig. 6.71. It follows that

$$R^{(1)}_{\text{reg}}(A = \infty_2)(\text{scan}_1(\sigma_1)) = -\delta\sigma_1.$$

Let us now consider $T = \sigma_1^2$. We will consider only $scan_2(T)$. There is again a unique closure to a circle. We show the regular isotopy $scan(\sigma_1^2)$ and the corresponding Gauss diagrams in Fig. 6.72 where we give names to some crossings too.

The case ∞_1:
The crossing c_2 is the only f-crossing for x_2 and the crossing hm is the only r-crossing for c_2. We have $\partial c_2 = \infty_1$, $W(x_2) = W_2(x_2) = 1$. The Reidemeister move x_3 is also a positive triple crossing with $\text{sign}(x_3) = -1$. We have $\partial(hm) = \partial(c_2) = \infty_1$ and $W_1(x_3) = 1$ (because of the crossing c_1). It follows that

$$R^{(1)}_{\text{reg}}(A = \partial\sigma_1^2)(\text{scan}(\sigma_1^2)) = 0$$

and that

$$R^{(1)}_{\text{reg}}(A = \infty_1)(\text{scan}(\sigma_1^2)) = R^{(1)}(\text{scan}(\sigma_1^2))$$

in this case. The calculation of $R^{(1)}(scan(\sigma_1^2))$ is given in Fig. 6.73. The result is

$$R^{(1)}_{\text{reg}}(\infty_1)(\text{scan}(\sigma_1^2)) = R^{(1)}(\text{scan}(\sigma_1^2)) = -\delta(1 + z.\sigma_1) = -\delta P_{\sigma_1^2}.$$

Here, as usual $\delta = (v - v^{-1})/z$ and $P_{\sigma_1^2}$ denotes the HOMFLYPT invariant of σ_1^2 in the skein module $S(\partial\sigma_1)$.

The case ∞_2:
The crossing c_1 is the only f-crossing for x_1 and there are no r-crossings. Consequently, $W(x_1) = 1$ and $W_2(x_1) = 0$. The crossing

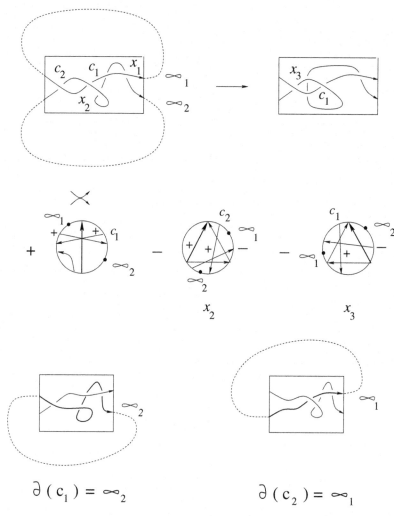

$$\partial\,(\,c_1\,) = \infty_2 \qquad\qquad \partial\,(\,c_2\,) = \infty_1$$

Figure 6.72:   The Reidemeister moves for $\mathrm{scan}(\sigma_1^2)$.

hm is the only f-crossing for x_2. We have $\partial(hm) = \partial(c_1) = \infty_2$, $W(x_2) = 1$ and $W_1(x_2) = W_2(x_2) = 0$ (because there are two r-crossings: A negative one and the crossing d which is positive). There is no r-crossing in x_3.

Consequently,

$$R_{\mathrm{reg}}^{(1)}(A = \partial\sigma_1^2)(\mathrm{scan}(\sigma_1^2)) = R^{(1)}(\mathrm{scan}(\sigma_1^2)) = 0.$$

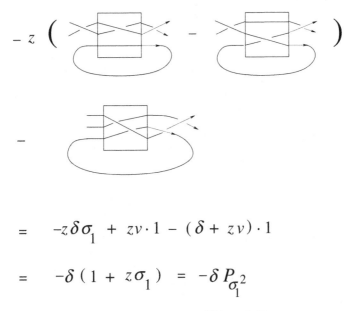

$$= \quad -z\delta\sigma_1 + zv \cdot 1 - (\delta + zv) \cdot 1$$

$$= \quad -\delta(1 + z\sigma_1) = -\delta P_{\sigma_1^2}$$

Figure 6.73: Calculation of $R^{(1)}(\text{scan}(\sigma_1^2))$ for ∞_1.

A similar calculation now gives

$$R^{(1)}_{\text{reg}}(A = \infty_2)(\text{scan}(\sigma_1^2)) = (-2v + v^{-1} - vz^2) \cdot \sigma_1.$$

For $T = \sigma_1$ and $T = \sigma_1^2$, we have $R^{(1)}(scan(T)) = 0$ for ∞_2. Therefore, we consider also the next case $T = \sigma_1^3$ for ∞_2. The Reidemeister moves in $scan(\sigma_1^3)$ are shown in Fig. 6.74. We have $W_2(x_1) = 1$, $W_1(x_2) = W_2(x_2) = 0$, $W_2(x_3) = 1$ (c_3 is an f-crossing which contributes with the r-crossing hm), $W_1(x_4) = W_2(x_4) = 1$ (c_1 is an r-crossing for hm, c_2 is an f-crossing but the corresponding r-crossings sum up to 0). The calculation of $R^{(1)}(scan(\sigma_1^3))$ is given in Fig. 6.75.

$$R^{(1)}(\text{scan}(\sigma_1^3)) = (-3v + 2v^{-1} - 4vz^2 + v^{-1}z^2 - vz^4) \cdot 1.$$

These simple examples show that $R^{(1)}_{\text{reg}}(A)$ depends on the grading A that both $R^{(1)}_{\text{reg}}(A)$ and $R^{(1)}$ depend on the choice of the point at infinity and of the choice of the curl. Moreover, we see that the two 1-cocycles are independent and that both are *not* multiplicative,

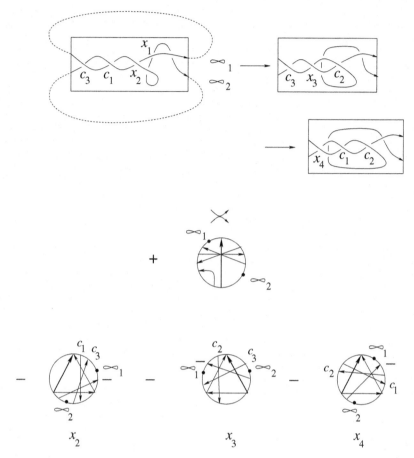

Figure 6.74: The Reidemeister moves for $\text{scan}(\sigma_1^3)$.

e.g., $R^{(1)}(\text{scan}(\sigma_1^2)) \neq (R^{(1)}(\text{scan}(\sigma_1)))^2$ for ∞_1 and $R^{(1)}(\text{scan}(\sigma_1^3)) \neq (R^{(1)}(\text{scan}(\sigma_1)))^3$ for ∞_2. Moreover, we see that $R^{(1)}(\text{scan}(\sigma_1^3))$ is *not* a multiple of $\delta P_{\sigma_1^3}$.

Example 6.2 We consider the very simple 3-tangle $T = \sigma_1\sigma_2$ with the standard braid closure and the scan-arc shown in Fig. 6.76. Let us chose the point $3 = \infty$. We calculate the version of $R_{\text{reg}}^{(1)}(A)$ and $R^{(1)}$ without contributions from II_0^-. Note that the crossing σ_1 has marking $\{2,3\}$, the crossing σ_2 has marking $\{3\}$ and these are the

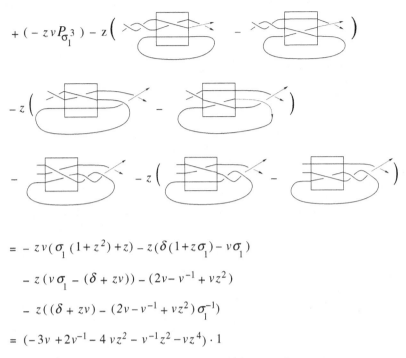

$$+ (-zvP_{\sigma_1^3}) - z\left(\quad - \quad \right)$$

$$-z\left(\quad - \quad \right)$$

$$- \quad -z\left(\quad - \quad \right)$$

$$= -zv(\sigma_1(1+z^2)+z) - z(\delta(1+z\sigma_1) - v\sigma_1)$$

$$-z(v\sigma_1 - (\delta + zv)) - (2v - v^{-1} + vz^2)$$

$$-z((\delta + zv) - (2v - v^{-1} + vz^2)\sigma_1^{-1})$$

$$= (-3v + 2v^{-1} - 4vz^2 - v^{-1}z^2 - vz^4) \cdot 1$$

Figure 6.75: Calculation of $R^{(1)}(\text{scan}(\sigma_1^3))$ for ∞_2.

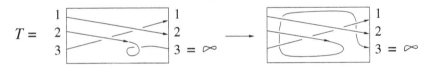

$$T = \quad \begin{array}{c} 1 \\ 2 \\ 3 \end{array} \qquad \begin{array}{c} 1 \\ 2 \\ 3 = \infty \end{array} \longrightarrow \begin{array}{c} 1 \\ 2 \\ 3 = \infty \end{array}$$

Figure 6.76: The scan-arc for $T = \sigma_1\sigma_2$.

only crossings with these markings. There are only two Reidemeister moves. They are positive Reidemeister III moves (i.e., type 1) of negative sign. Their Gauss diagrams are shown in Fig. 6.77. One easily sees that $R^{(1)}(\text{scan}(T)) = 0$ because all weights $W_1 = W_2 = 0$. Moreover, $R^{(1)}_{\text{reg}}(\{2,3\})(\text{scan}(T)) = R^{(1)}_{\text{reg}}(\{3\})(\text{scan}(T))$ because the partial smoothings T_{r_a} for x_1 and x_2 lead to identical diagrams and the partial smoothings $T(\text{type})$ for x_1 and x_2 enter into both invariants.

The calculation of $R^{(1)}_{\text{reg}}(\{2,3\})(\text{scan}(T))$ is given in Fig. 6.78.

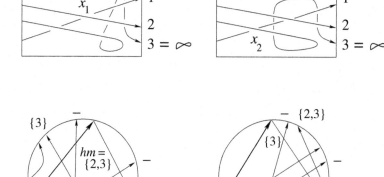

Figure 6.77:   The Gauss diagrams in the scan-arc.

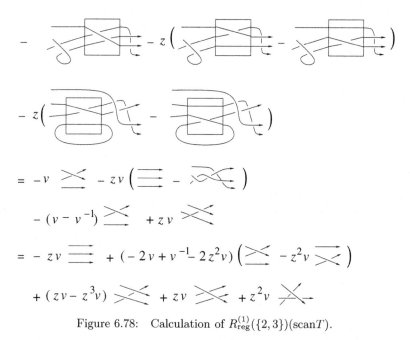

Figure 6.78:   Calculation of $R_{\text{reg}}^{(1)}(\{2,3\})(\text{scan}T)$.

We see that even for this extremely simple tangle T (just a basis element of the skein module of 3-braids) our new invariants are already rather complex, i.e., they contain lots of monomials and in particular all six basis elements of the skein module of 3-braids occur in the value of the invariant.

6.9   The 1-Cocycle $R^{(1)}$ in the Complement of Cusps with a Transverse Branch

We know already that $R^{(1)}$ is a 1-cocycle in M_T^{reg}. First of all in M_T we normalise the HOMFLYPT invariants as usual by $v^{-w(T)}$, where $w(T)$ is the writhe of T. Moreover, we fix an (abstract) closure σ of T and a point at infinity ∞ in ∂T (in the case of long knots, there is only a canonical choice).

There are several local types of Reidemeister I moves: the new crossing could be positive or negative. There are also several global types of Reidemeister I moves: the new crossing could be of type 0 or 1.

Definition 6.22 *The partial smoothing $T_I(p)$ of a Reidemeister I move p of global type 0 is defined by*

$$T_I(p) = -1/2(v - v^{-1})\delta v_2(T)P_T.$$

Here, $\delta = (v - v^{-1})/z$ is the HOMFLYPT polynomial of the triv- ial link of two components and P_T denotes the (normalised) HOM- FLYPT invariant of T in the skein module $S(\partial T)$. The integer $v_2(T)$ for a given (abstract) closure σ of T and a given choice of ∞ in ∂T is defined by the first Polyak–Viro formula in Fig. 1.3 with ∞ as the marked point.

The partial smoothing $T_I(p)$ of a Reidemeister I move p of global type 1 is 0.

We now add $\text{sign}(p)T_I(p)$ to $R^{(1)}$ for each Reidemeister I move p.

If T is a long knot, then $v_2(T)$ is just the Vassiliev invariant of degree 2 of T (which is, e.g., equal to the coefficient of z^2 in the Conway polynomial of T, see, e.g., Ref. [2]).

So, Reidemeister I moves of global type 1 do not contribute to the 1-cocycle $R^{(1)}$. Note that the contribution of a Reidemeister I move of global type 0 does not depend on the local type neither on its place in the diagram T. Hence, in particular it has the scan-property.

Let s be a loop in M_T. It is not difficult to see that the number of Reidemeister I moves of global type 0 is always even (in fact, the algebraic number of Reidemeister I moves of global type 0 which are of positive local type is always equal to the algebraic number of Reidemeister I moves of global type 0 which are of negative local type because they always come in pairs in Reidemeister II moves).

Lemma 6.7 *The value of the 1-cocycle $R^{(1)}$ on a meridian of $\Sigma^{(1)} \cap \Sigma^{(1)}_{\text{cusp}}$ is zero.*

Proof. The new crossing from $\Sigma^{(1)}_{\text{cusp}}$ could be an f-crossing for the Reidemeister move in $\Sigma^{(1)}$. However, it is an isolated crossing and in particular it has no r-crossings at all. The changing for the HOMFLYPT invariants for $\Sigma^{(1)}$, which comes from the new crossing in $\Sigma^{(1)}_{\text{cusp}}$, is compensated by the normalisation. The contribution of $\Sigma^{(1)}_{\text{cusp}}$ doesn't change at all under isotopy of the tangle outside the cusp. □

We now have to deal with the irreducible strata of codimension two which contain a diagram with a cusp. There are exactly 16 different local types of strata corresponding to a cusp with a transverse branch. We list them in Figs. 6.80–6.83, where we move the branch from the right-hand side to the left-hand side. For each local type, we have exactly two global types. We give in the figure also the Gauss diagrams of the triple crossing and of the self-tangency with equal tangent direction. Note that in each Gauss diagram of a triple crossing, one of the three arcs is empty besides just one head or foot of an arrow. Let us denote each stratum of $\Sigma^{(2)}_{\text{trans}-\text{cusp}}$ simply by the global type of the corresponding triple crossing.

Lemma 6.8 *The value of the 1-cocycle $R^{(1)}$ on a meridian is zero for all strata in $\Sigma^{(2)}_{\text{trans}-\text{cusp}}$ besides those in $\Sigma^{(2)}_{l_c}$.*

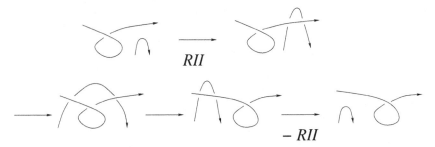

Figure 6.79:   Two Reidemeister II moves with cancelling contributions.

Proof. It is clear that we have to consider only one orientation of the moving branch because the contributions of the two Reidemeister II moves in Fig. 6.79 always cancel out, no matter the orientations, the types and whether the branches move over or under the cusp. We show first that $R^{(1)}$ vanishes on all Whitney tricks. It follows then from Figs. 6.80–6.83 that it suffices to prove the assertion in four cases: l_b of type 1 and the branch moves over the cusp, l_b of type 1 and the branch moves under the cusp, r_b of type 1 and r_a of type 1. Note that the Reidemeister I move in the meridian does never contribute because it enters twice but with opposite signs.

The first Whitney trick is shown in Fig. 6.84 and the calculation in Fig. 6.85. We write K for T and one easily sees that all weights W_2 are equal to v_2. The second Whitney trick is rather similar to the first, see Figs. 6.86 and 6.87. In the third and fourth Whitney tricks all distinguished crossings d are of type 1. It suffices to establish that the triple crossing is not of type l_c. We do this in Figs. 6.88 and 6.89. Consequently, there are no contributions at all for these Whitney tricks.

Let us consider now the above four cases. We study the Gauss diagrams in detail for the case l_b with the branch moving over the cusp. The Reidemeister moves with the Gauss diagrams are given in Fig. 6.90. Note that the self-tangency with opposite tangent direction and the triple crossing share the same distinguished crossing d. The distinguished crossing of the other self-tangency has almost the same head as d but not the same foot. However, their foots are connected by an empty arc in the circle. There are exactly two sorts

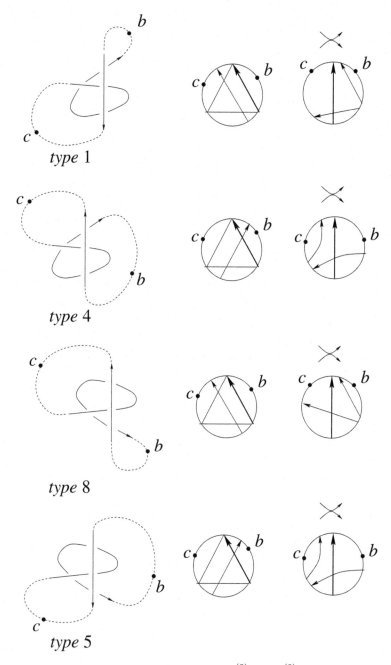

Figure 6.80:   The strata $\Sigma_{l_c}^{(2)}$ and $\Sigma_{l_b}^{(2)}$.

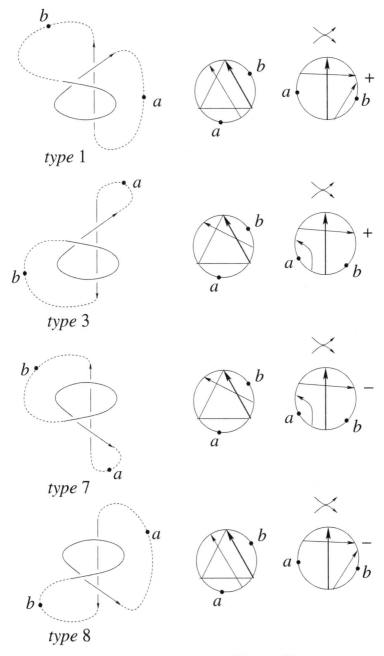

type 1

type 3

type 7

type 8

Figure 6.81: The strata $\Sigma^{(2)}_{l_a}$ and $\Sigma^{(2)}_{l_b}$.

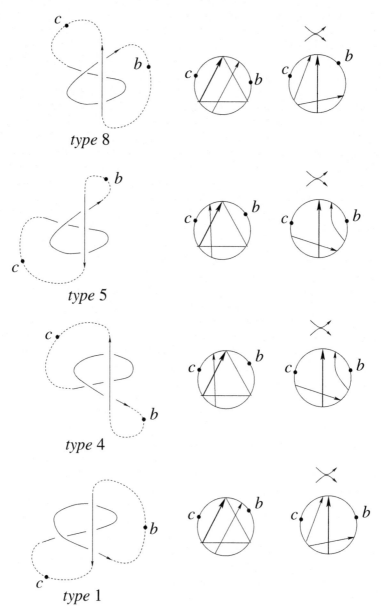

type 8

type 5

type 4

type 1

Figure 6.82:   The strata $\Sigma_{r_b}^{(2)}$ and $\Sigma_{r_c}^{(2)}$.

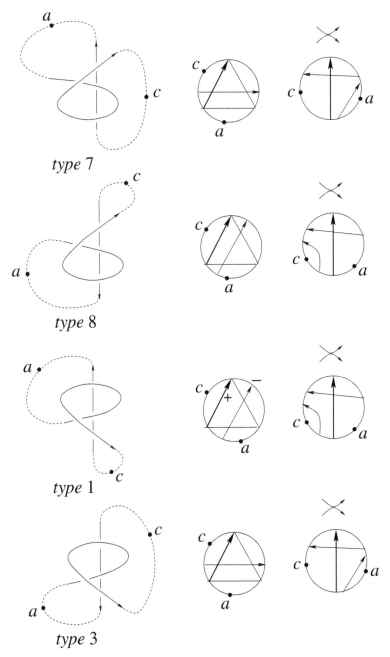

Figure 6.83: The strata $\Sigma_{r_a}^{(2)}$ and $\Sigma_{r_c}^{(2)}$.

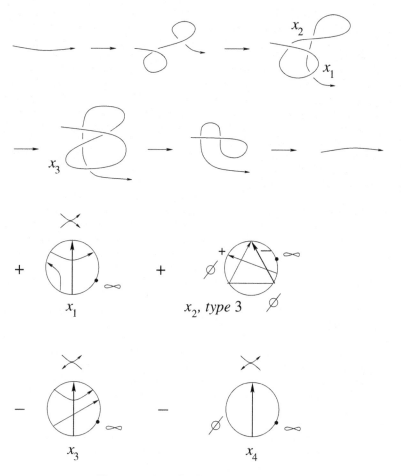

Figure 6.84:   The first Whitney trick.

of f-crossings, which we call f_1 and f_2 (compare Fig. 6.90). In all three Gauss diagrams, each corresponding f_1-crossing shares the same r-crossing. In x_1, each f_2-crossing has already exactly one positive r-crossing in the drawn part of the Gauss diagram. There are three r-crossings in x_2 but their sum is $+1$. In x_3, there is again a single positive r-crossing (but different from that in x_1). All other r-crossings (not drawn in the diagrams) are the same for each corresponding f_2-crossing. Consequently, we have shown that the weights W_2 are the same for all three Reidemeister moves. We now give the calculation of $R^{(1)}$ in Fig. 6.91.

$$- 1/2 \; v_2 \; (v - v^{-1}) \, \delta \, P_K \; - \; 1/2 \; v_2 \; (v - v^{-1}) \, \delta \, P_K$$

$$- v_2 \, z \, P_K \; + \; v_2 \, z \, \left(\; \text{⬡} \; - \; \text{⬡} \; \right)$$

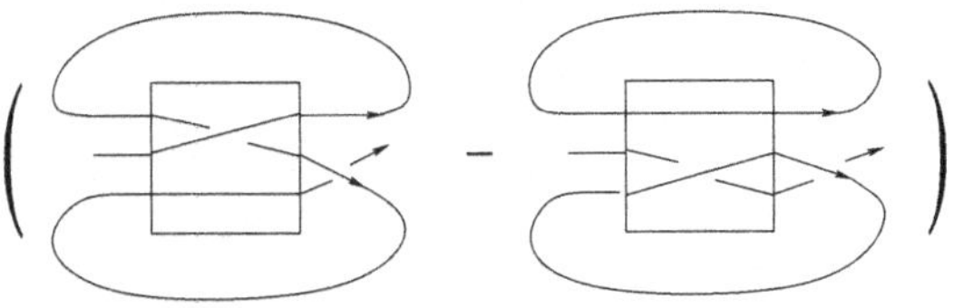

$$+ \; v_2 \; (v - v^{-1}) \; \text{⟲} \; + \; v_2 \; (v - v^{-1}) \; \text{⟲} \; = \; 0$$

Figure 6.85: Calculation of $R^{(1)}$ for the first Whitney trick.

In exactly the same way one shows that the weights W_2 of the three moves are the same in each of the remaining three cases. Therefore, we give only the calculations of the sum of the partial smoothings in Figs. 6.92–6.94. Note that in the last case the triple crossing could also contribute with a partial smoothing associated with W_1. However, we see immediately from Fig. 6.83 that $W_1 = 0$. Indeed, there are only exactly two crossings which contribute to W_1 because there is only exactly one head of an arrow in the arc in the circle from the middle crossing to the upper crossing. But the two crossings have opposite signs (which are indicated in the figure too). □

Lemma 6.9 *The value of the 1-cocycle $R^{(1)}$ on a meridian of a degenerate cusp, locally given by $x^2 = y^5$ and denoted by $\Sigma^{(2)}_{|rmcusp-deg}$, is zero.*

Proof. Again, the cusp can be of type 0 or of type 1. We show the meridian of the degenerate cusp of type 0 in Fig. 6.95 and the calculation in Fig. 6.96. Note that $W_2(p) = v_2(T)$ for the self-tangency p. The meridian of the degenerate cusp of type 1 is shown in Fig. 6.97. In this case, again no Reidemeister move at all contributes to $R^{(1)}$. □

We have finally shown that $R^{(1)}(m) = 0$ for the meridians m of all six types of strata of codimension 2 (given in Section 6.2) besides for

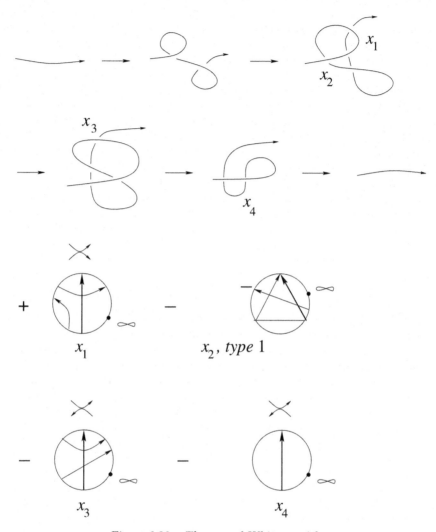

Figure 6.86:   The second Whitney trick.

the four local types of strata from $\Sigma_{l_c}^{(2)}$. Theorem 6.2 implies that $R^{(1)}$ has the scan-property. Consequently, we have proven the following proposition.

$$- 1/2 \; v_2 \, (v - v^{-1}) \, \delta \, P_K \; - \; 1/2 \; v_2 \, (v - v^{-1}) \, \delta \, P_K$$

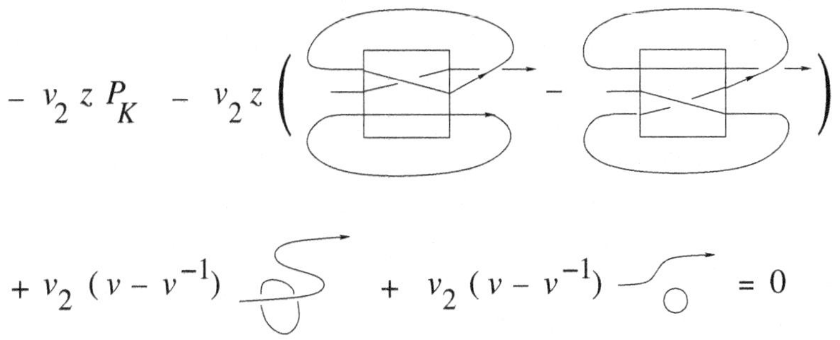

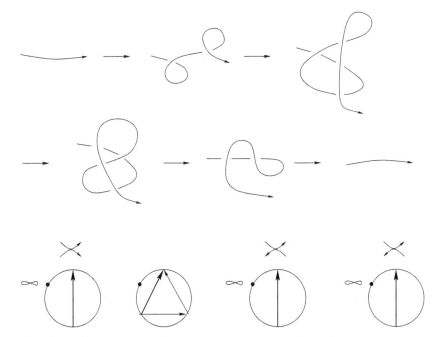

$$- \; v_2 \, z \, P_K \; - \; v_2 \, z \left(\rule{0pt}{40pt} \hspace{180pt} - \hspace{120pt} \right)$$

$$+ \; v_2 \, (v - v^{-1}) \hspace{50pt} + \; v_2 \, (v - v^{-1}) \hspace{50pt} = 0$$

Figure 6.87: Calculation of $R^{(1)}$ for the second Whitney trick.

Figure 6.88: The third Whitney trick.

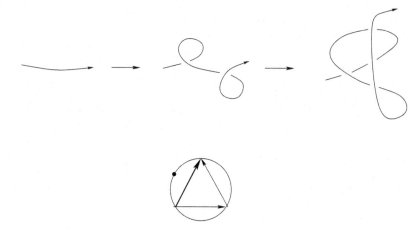

Figure 6.89:   The fourth Whitney trick.

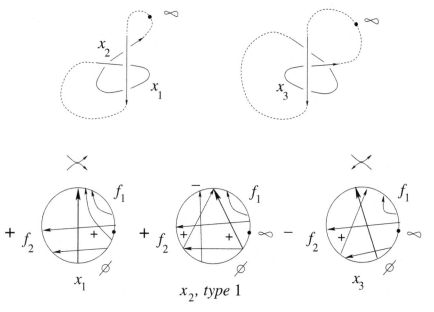

Figure 6.90:   The meridian for $\Sigma_{l_b}^{(2)}$ of local type 1 with over branch.

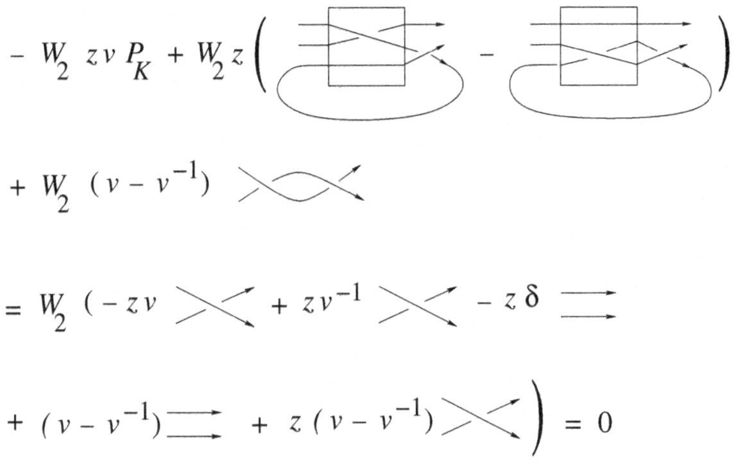

$$- W_2 \, zv \, P_K + W_2 \, z \left(\begin{array}{c} \end{array} - \begin{array}{c} \end{array} \right)$$

$$+ W_2 \, (v - v^{-1}) \begin{array}{c} \end{array}$$

$$= W_2 \left(-zv \begin{array}{c} \end{array} + zv^{-1} \begin{array}{c} \end{array} - z\delta \begin{array}{c} \end{array} \right.$$

$$+ (v - v^{-1}) \begin{array}{c} \end{array} + z(v - v^{-1}) \begin{array}{c} \end{array} \left. \right) = 0$$

Figure 6.91: Calculation for $\Sigma^{(2)}_{l_b}$ of local type 1 with over branch.

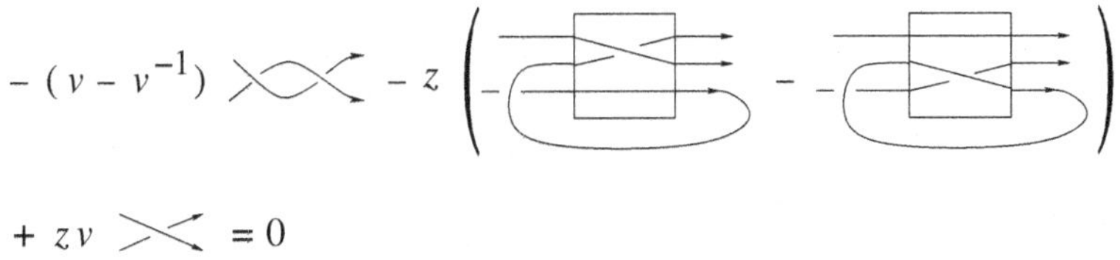

$$- (v - v^{-1}) \begin{array}{c} \end{array} - z \left(\begin{array}{c} \end{array} - \begin{array}{c} \end{array} \right)$$

$$+ zv \begin{array}{c} \end{array} = 0$$

Figure 6.92: Calculation for $\Sigma^{(2)}_{l_b}$ of local type 1 with under branch.

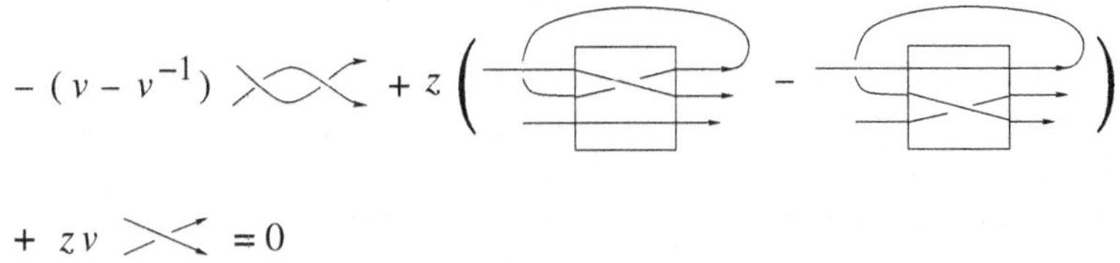

$$- (v - v^{-1}) \begin{array}{c} \end{array} + z \left(\begin{array}{c} \end{array} - \begin{array}{c} \end{array} \right)$$

$$+ zv \begin{array}{c} \end{array} = 0$$

Figure 6.93: Calculation for $\Sigma^{(2)}_{r_b}$ of local type 1.

Proposition 6.6 *Let s be a generic oriented arc in M_T (with a fixed abstract closure $T \cup \sigma$ to an oriented circle and a fixed point at infinity*

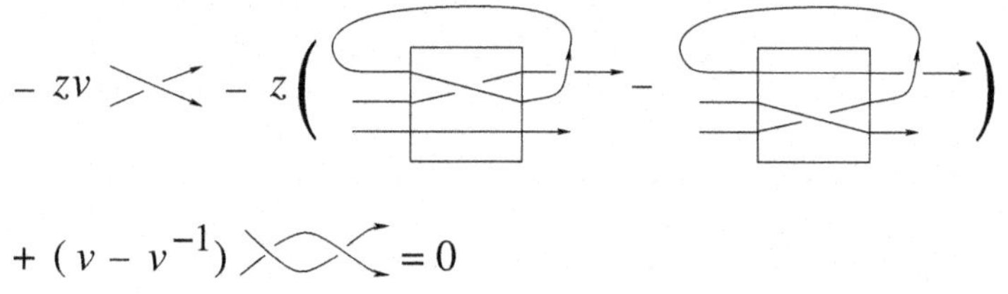

$$- zv \; \times\!\!\!<\;\; - \; z \left(\qquad\qquad\qquad \rightarrow \; - \qquad\qquad\qquad \rightarrow \right)$$

$$+ \; (v - v^{-1}) \; \times\!\!\!\!\times\!\!\!< \; = 0$$

Figure 6.94:   Calculation for $\Sigma_{r_a}^{(2)}$ of local type 1.

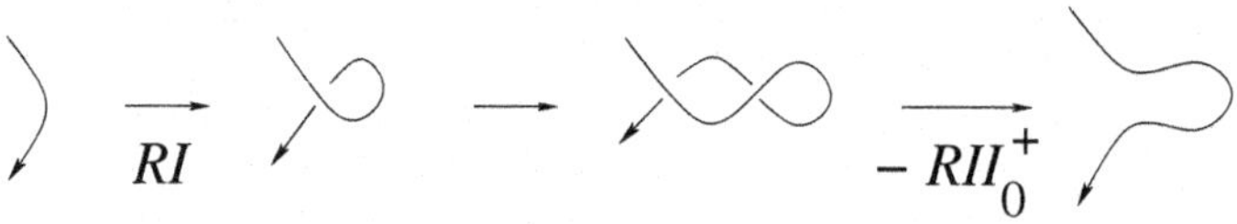

$$) \; \xrightarrow{\;RI\;} \; \text{)O} \; - \; \text{)OO} \; \xrightarrow{\; -RII_0^+ \;} \; \text{)}$$

Figure 6.95:   Meridian of a degenerated cusp of type 0.

$$- 1/2 \, (v - v^{-1}) v_2 (T) \delta P_T \; - 1/2 \, (v - v^{-1}) v_2 (T) \delta P_T$$

$$- (- (v - v^{-1}) v_2 (T) \delta P_T \,) = 0$$

Figure 6.96:   Calculation of $R^{(1)}$ on the meridian of a degenerated cusp of type 0.

$$) \; \xrightarrow{\;RI\;} \; \text{)O} \; \xrightarrow{\;RI\;} \; \text{)OO} \; \xrightarrow{\; -RII_1^+ \;} \; \text{)}$$

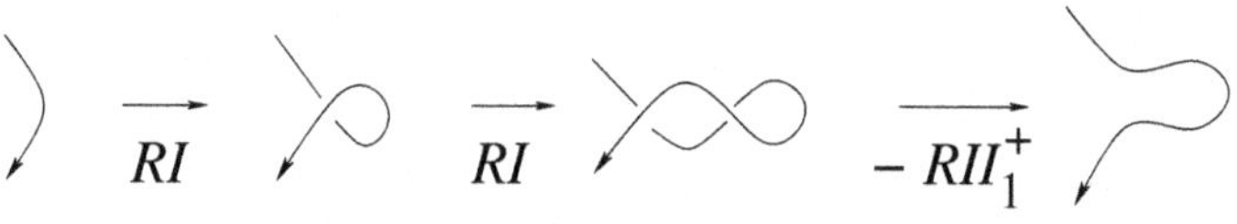

Figure 6.97:   Meridian of a degenerated cusp of type 1.

in ∂T). Then, the 1-cochain

$$R^{(1)}(s) = \sum_{p \in s \cap l_c} \operatorname{sign}(p) W_1(p) T_{l_c}(\text{type})(p)$$

$$+ \sum_{p \in s \cap r_a} \operatorname{sign}(p) [W_1(\text{type})(p) T_{r_a}(\text{type})(p)$$

$$+ z W_2(\text{type})(p) T(\text{type})(p)]$$

$$+ \sum_{p \in s \cap r_b} \operatorname{sign}(p) z W_2(p) T(\text{type})(p)$$

$$+ \sum_{p \in s \cap l_b} \text{sign}(p) z W_2(p) T(\text{type})(p)$$

$$+ \sum_{p \in s \cap II_0^+} \text{sign}(p) W_2(p) T_{II}(p)$$

$$+ \sum_{p \in s \cap II_0^-} \text{sign}(p) W_2(p) z P_T$$

$$+ \sum_{p \in s \cap I} \text{sign}(p) T_I(p)$$

is a 1-cocycle in $M_T \backslash \Sigma_{l_c}^{(2)}$ *and it has the scan-property for branches moving under the tangle.*

Lemma 6.10 *The value of* $R^{(1)}$ *on the meridian of* $\Sigma_{l_c}^{(2)}$ *is equal to* $+W_1(p)\delta P_T$ *or* $-W_1(p)\delta P_T$ *as shown in Fig. 6.98.*

Proof. We observe that the fourth involved crossing, which is not in the triangle, does not contribute to $W_1(p)$. The rest is a straightforward calculation. Note that the local types 8 and 5 force us to use the version of $R^{(1)}$ which associates $\text{sign}(p)W_2(p)zP_T$ to Reidemeister moves of type II_0^- instead of the more complicated partial smoothings for the global types r_a and l_c in the case of the local types $2, 3, 5, 8$ and no contribution from II_0^- at all (compare Remark 6.9). □

6.10 The Finite Type 1-Cocycle $V^{(1)}$ in the Complement of Cusps with a Transverse Branch

Let us summarise: the quantum part $R^{(1)}$ of the 1-cocycle uses the global types r_a, r_b, l_b, l_c of Reidemeister III moves and the types II_0^+ and II_0^- of Reidemeister II moves. Note that in all cases the distinguished crossing d is of type 0 besides in the case l_c. The value of $R^{(1)}$ on the meridian of $\Sigma_{\text{trans-cusp}}^{(2)}$ is non-zero exactly in the latter case. This happens because the distinguished crossing d is of type 1 for the Reidemeister II moves in the corresponding unfolding and it does not compensate the contribution of the triple crossing. This forces us to consider the moves of type II_1^- too.

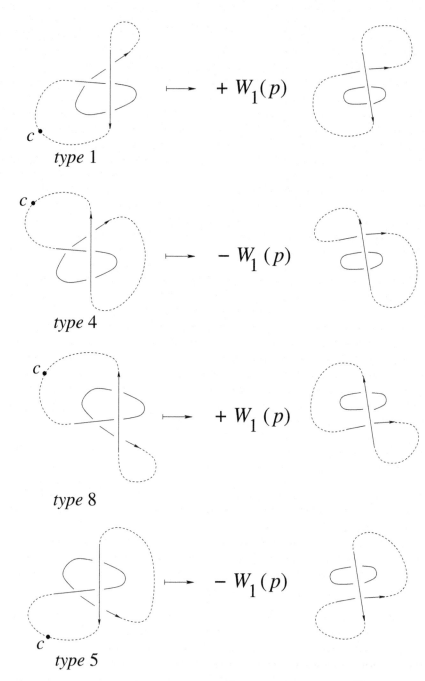

Figure 6.98: The values of $R^{(1)}$ on the meridians of $\Sigma_{l_c}^{(2)}$.

Definition 6.23 *The weight $W(p)$ of a Reidemeister II move of type II_1^- is defined as the sum of the writhes $w(q)$ of all crossings q of the configuration given in Fig. 6.99.*

Remember that we simply say that the weight is defined by the configuration in the figure (compare Section 6.3).

Lemma 6.11 *The weight $W(p)$ of a Reidemeister II move of type II_1^- satisfies the r-cube equations.*

Proof. This follows immediately from the inspection of the figures of the r-cube equations in Section 6.6.   □

However, this weight doesn't satisfy the l-cube equations. Therefore, we have to associate a weight with the remaining global types of Reidemeister III moves too.

Definition 6.24 *The weight $W(d)$ of a Reidemeister III move of type r_c is defined by the configurations in Fig. 6.100.*

Definition 6.25 *The preliminary weight W_1 of a Reidemeister III move of type l_a is defined by the configurations in Fig. 6.101.*

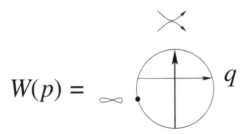

Figure 6.99:   The weight $W(p)$ for Reidemeister moves of type II_1^-.

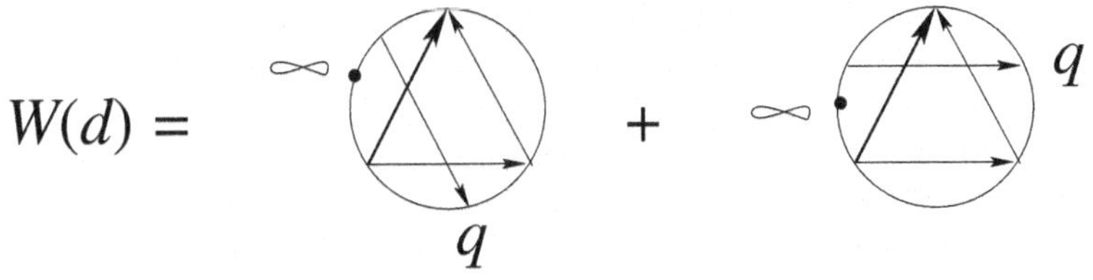

$$W(d) = \quad + \quad$$

Figure 6.100:   The weight $W(d)$ of a Reidemeister III move of type r_c.

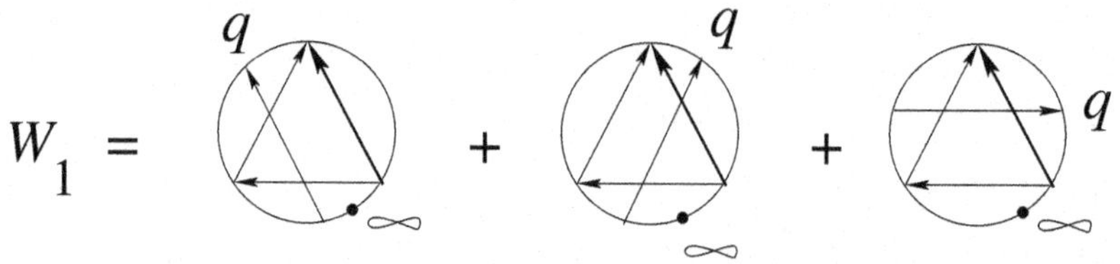

$$W_1 = \quad + \quad + \quad q$$

Figure 6.101: The preliminary weight W_1 of a Reidemeister III move of type l_a.

The weight $W_1(\text{type})$ depends moreover on the local type of the Reidemeister III move and is given in the following:

$$W_1(\text{type1}) = W_1(\text{type3}) = W_1(\text{type6}) = W_1 + 1$$
$$W_1(\text{type2}) = W_1(\text{type7}) = W_1(\text{type8}) = W_1 - 1$$
$$W_1(\text{type4}) = W_1(\text{type5}) = W_1.$$

We are now ready to define the 1-cochain $V^{(1)}$ (the letter "V" stands for Vassiliev).

Definition 6.26 *Let s be a generic oriented arc in M_T (with a fixed abstract closure $T \cup \sigma$ to an oriented circle and a fixed point at infinity in ∂T). Then, the 1-cochain $V^{(1)}$ is defined as*

$$V^{(1)}(s) = - \sum_{p \in s \cap l_a} \text{sign}(p) W_1(\text{type}) + \sum_{p \in s \cap r_c} \text{sign}(p) W(d)$$
$$+ \sum_{p \in s \cap II_1^-} \text{sign}(p) W(p).$$

Lemma 6.12 *$V^{(1)}$ satisfies the positive global tetrahedron equation. Moreover, the contribution of $-P_2 + \bar{P}_2$ is always zero.*

Proof. First of all, we observe that we can forget about the constant correction $W_1(\text{type } 1) = W_1 + 1$ because each stratum appears twice and with different signs. We inspect the figures in Section 6.3 and we sum up the contributions $V^{(1)}$ from $-P_1$ to $-\bar{P}_4$:

I_1: 0, I_2: 0, I_3: 0, I_4: $-2 + 2$.

II_1: 0, II_2: $+1 - 1$, II_3: 0, II_4: 0.

III_1: 0, III_2: 0, III_3: $-1 - 1 - 1 + 2 + 1$, III_4: 0.

IV_1: 0, IV_2: $+1 - 1 - 1 + 1$, IV_3: 0, IV_4: 0.

V_1: 0, V_2: $+1 - 1$, V_3: 0, V_4: $+2 - 2$.
VI_1: 0, VI_2: 0, VI_3: $+1 - 2 - 2 + 1 + 2$, VI_4: 0.

Moreover, we see that the contributions of $-P_2$ and $+\bar{P}_2$ always cancel out. ☐

Lemma 6.13 $V^{(1)}$ *vanishes on the meridians of* $\Sigma^{(1)} \cap \Sigma^{(1)}$, $\Sigma^{(2)}_{\text{self}-\text{flex}}$ *and* $\Sigma^{(2)}_{\text{cusp}-\text{deg}}$.

Proof. Completely obvious. But note that the strata $\Sigma^{(2)}_{\text{cusp}-\text{deg}}$ force us to associate $\text{sign}(p)W(p)$ with Reidemeister moves of type II_1^- instead of type II_1^+. ☐

Lemma 6.14 $V^{(1)}$ *is zero on all meridians of* $\Sigma^{(2)}_{\text{trans}-\text{cusp}} \setminus \Sigma^{(2)}_{l_c}$.

Proof. By inspection of Figs. 6.80–6.83 from the previous section, note that the four cases with a triple crossing of global type l_a force us to define the above constant corrections for the local types 1, 3, 7 and 8. Note that this correction term is exactly $w(hm)$ for these four cases. The correction term for the remaining cases is forced by the cube equations. ☐

Lemma 6.15 $V^{(1)}$ *satisfies the cube equations.*

Proof. We know already that the weight $W(p)$ of a Reidemeister II move of type II_1^- satisfies the r-cube equations (Lemma 6.11). The weight $W(d)$ of a Reidemeister III move of type r_c satisfies evidently the r-cube equations because the arrow q cannot slide onto an arrow of the triangle without sliding over ∞ (compare Observation 6.1 in Section 6.6). So, we are left with the l-cube equations for $\infty = a$. We inspect the figures from Section 6.6 (and of course the signs from Fig. 2.2 and from Definition 6.4). The contribution of the triple crossings is always in brackets (do not forget about the constant correction term).

"1-7": $-(-1+1) = -(+1-1)$, "1-5": $-(0+1) = -1-(0)$, "5-3": $-(0)-1 = -(0+1)$, "7-4": $(0-1)+1 = (0)$, "3-8": $-(-1+1) = -(+1-1)$, "4-8": $(0) = +1+(0-1)$.

We determine now the correction term for the star-like triple crossings. Reidemeister moves of type II_1^- do no longer occur.

"1–6": $-(0+1) = -(0+1)$, "7–2": $-(0-1) = -(0-1)$, "3-6": $(0+1) = (0+1)$, "4–6": $(0) = (-1+1)$, "5–2": $(0) = (+1-1)$, "8–2": $-(0-1) = -(0-1)$. ☐

The following remark is very important.

Remark 6.8 *It is tempting to complete $V^{(1)}$ to a 1-cocycle in M_T by adding $\sum_{p \in s \cap l_c} \mathrm{sign}(p) W_1(p)$ (compare Definition 6.12). Then, it would indeed be zero on all meridians of $\Sigma^{(2)}_{\mathrm{trans-cusp}}$. However, $\sum_{p \in s \cap l_c} \mathrm{sign}(p) W_1(p)$ does not satisfy the global positive tetrahedron equation as one sees immediately from the figures in Section 6.3 for the global case II and $\infty = 2$. Hence, the quantum part $R^{(1)}$ of the 1-cocycle is essential in order to obtain a 1-cocycle for the whole M_T.*

Proposition 6.7 $V^{(1)}$ *is an integer valued 1-cocycle in $M_T \backslash \Sigma^{(2)}_{l_c}$. Its value on a meridian of $\Sigma^{(2)}_{l_c}$ is equal to $+W_1(p)$ or $-W_1(p)$ of the triple crossing. It has the scan-property for branches moving under the tangle T.*

Proof. It follows from the lemmas in this section that $V^{(1)}$ vanishes on the meridians of five of the six types of strata of codimension two (compare Section 6.2). Moreover, according to Lemma 6.14, it vanishes also on the meridians of $\Sigma^{(2)}_{\mathrm{trans-cusp}} \backslash \Sigma^{(2)}_{l_c}$. It follows that $V^{(1)}$ is a 1-cocycle in $M_T \backslash \Sigma^{(2)}_{l_c}$. It has the scan-property for positive triple crossings according to Lemma 6.12. Inspecting the figures in Section 6.6, we see that it has the scan-property for Reidemeister moves of type II_1^- (and hence for all Reidemeister moves of type II). It follows now that $V^{(1)}$ has the scan-property because the graph Γ is connected. This is exactly the same argument like in the case of $R^{(1)}$. ☐

Remark 6.9 $V^{(1)}$ *cannot be defined as a 1-cocycle of finite type in Vassiliev's [46] sense because it is only defined in $M_T \backslash \Sigma^{(2)}_{l_c}$ (we just take out something of codimension two in the components of smooth, non-singular knots and we will not see anything dual in the discriminant of singular knots). However, it is of finite type in the sense that it can be given by a Gauss diagram formula of finite type (i.e., a finite number of arrows in each configuration in the formula). We define the Gauss degree of a Gauss diagram formula in the last section. According to this definition, $V^{(1)}$ is of Gauss degree 3.*

6.11 The 1-Cocycle $\bar{R}^{(1)} = R^{(1)} - \delta P_T V^{(1)}$ Represents a Non-Trivial Cohomology Class in the Topological Moduli Space

Definition 6.27 *The completion* $\bar{R}^{(1)}$ *is defined by* $\bar{R}^{(1)} = R^{(1)} - \delta P_T V^{(1)}$.

The following theorem is the most important result in this chapter.

Theorem 6.3 *Let s be a generic oriented arc in M_T (with a fixed abstract closure $T \cup \sigma$ to an oriented circle and a fixed point at infinity in ∂T). Then, the 1-cochain with values in $S(\partial T)$.*

$$\bar{R}^{(1)}(s) = R^{(1)}(s) - \delta P_T V^{(1)}(s)$$

$$= + \sum_{p \in s \cap l_c} \text{sign}(p) W_1(p) T_{l_c}(\text{type})(p)$$

$$+ \sum_{p \in s \cap r_a} \text{sign}(p) [W_1(\text{type})(p) T_{r_a}(\text{type})(p)$$

$$+ z W_2(\text{type})(p) T(\text{type})(p)]$$

$$+ \sum_{p \in s \cap r_b} \text{sign}(p) z W_2(p) T(\text{type})(p)$$

$$+ \sum_{p \in s \cap l_b} \text{sign}(p) z W_2(p) T(\text{type})(p)$$

$$+ \sum_{p \in s \cap II_0^+} \text{sign}(p) W_2(p) T_{II}(p) + \sum_{p \in s \cap II_0^-} \text{sign}(p) W_2(p) z P_T$$

$$+ \sum_{p \in s \cap I} \text{sign}(p) T_I(p)$$

$$+ \sum_{p \in s \cap l_a} \text{sign}(p) W_1(\text{type})(p) \delta P_T - \sum_{p \in s \cap r_c} \text{sign}(p) W(p) \delta P_T$$

$$- \sum_{p \in s \cap II_1^-} \text{sign}(p) W(p) \delta P_T$$

is a 1-cocycle in M_T. It represents a non-trivial cohomology class and it has the scan-property for branches moving under the tangle T.

For the convenience of the reader, all the weights are summarised in Fig. 6.102. Here, d is the distinguished crossing of an R III move (i.e., the crossing between the highest and the lowest branch) or the two new crossings together of an R II move or the new crossing of an R I move. Remember our convention that the contribution of a

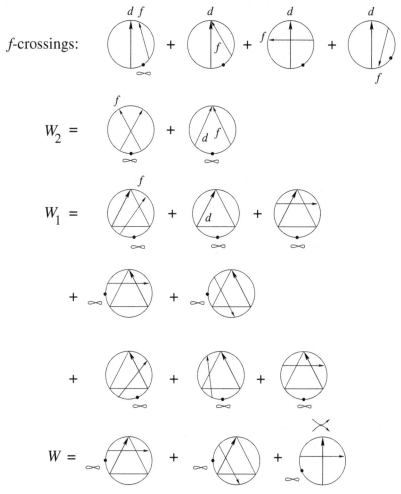

Figure 6.102:   The weights for $\bar{R}^{(1)}$.

configuration (i.e., a Gauss diagram without writhes) is the sum of the product of the writhe of the arrows (not in the triangle). Hence, W_1 and W are linear weights and W_2 is a quadratic weight. The partial smoothings are summarised in Figs. 6.103–6.111. Here, we show the triple crossing and the self-tangency on the negative side of the discriminant $\Sigma^{(1)}$, i.e., the Reidemeister move to the other side of the discriminant enters with the sign $+1$. The distinguished crossing d is always drawn with a thicker arrow. Remember also that δ is the HOMFLYPT polynomial of the trivial link of two components, that $v_2(T)$ is the invariant of degree two defined by the first Polyak–Viro

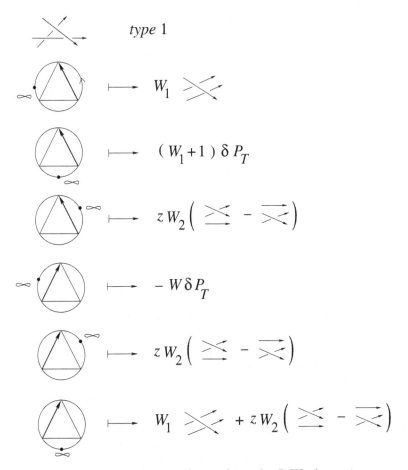

Figure 6.103: The partial smoothings for R III of type 1.

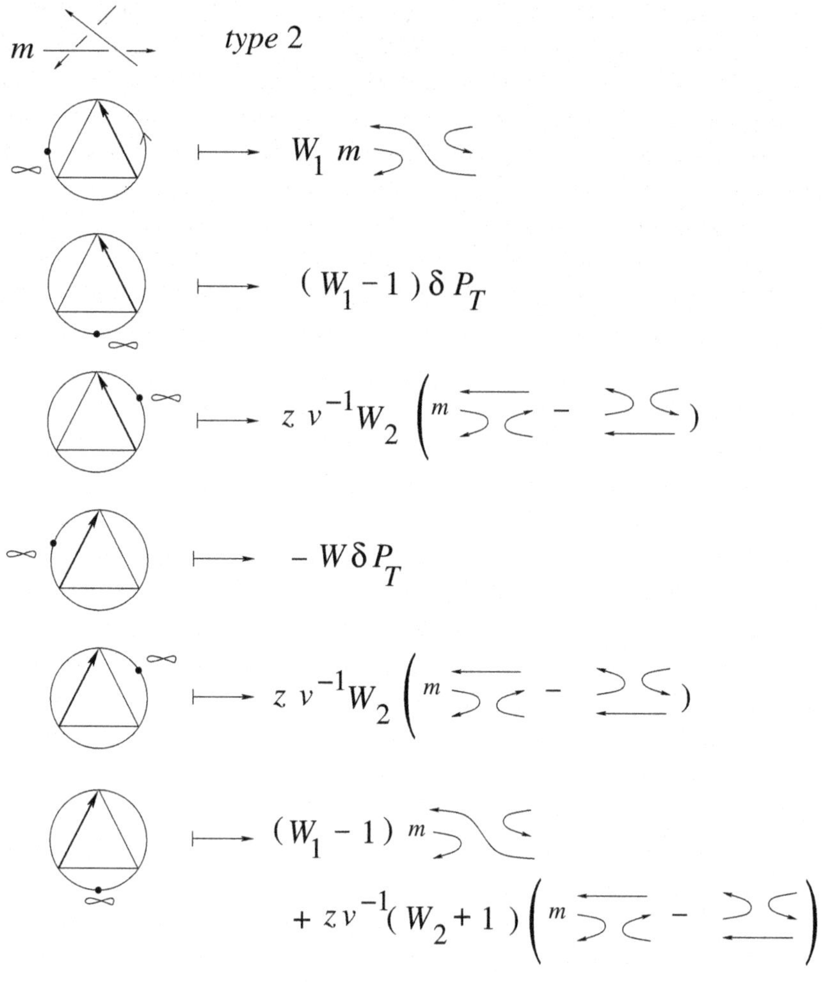

$$m \longrightarrow \text{type 2}$$

$\longmapsto W_1 m$

$\longmapsto (W_1 - 1) \delta P_T$

$\longmapsto z\, v^{-1} W_2 \left(m \;\; - \;\; \right)$

$\longmapsto -W \delta P_T$

$\longmapsto z\, v^{-1} W_2 \left(m \;\; - \;\; \right)$

$\longmapsto (W_1 - 1)\, m$

$$+ z v^{-1}(W_2 + 1) \left(m \;\; - \;\; \right)$$

Figure 6.104:   The partial smoothings for R III of type 2.

formula in Fig. 1.3 and that P_T is the element in the HOMFLYPT skein module $S(\partial T)$ represented by T.

Proof.   It follows from Lemma 6.6 and Propositions 6.6 and 6.7 that $\bar{R}^{(1)}$ is a 1-cocycle in $M_T \backslash \Sigma_{l_c}^{(2)}$ and that it has the scan-property. It remains to show that it vanishes on the meridians of $\Sigma_{l_c}^{(2)}$. But we see this immediately from Fig. 6.98 and from the

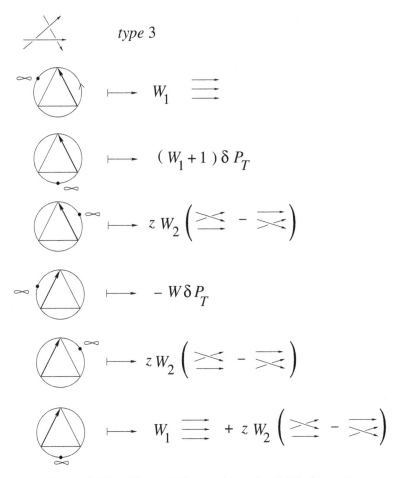

Figure 6.105: The partial smoothings for R III of type 3.

corresponding Gauss diagrams in Fig. 6.80 (compare also the proof of Lemma 6.8). □

An important property of this 1-cocycle is the fact that it represents a non-trivial cohomology class in particular for $\amalg_K M_K$, where K runs over all isotopy classes of long knots.

Example 6.3 We start with showing that $\bar{R}^{(1)}(rotK) = -\delta P_K$ for K the right trefoil (compare the Introduction). A version of the loop $rot(K)$ as almost regular isotopy is shown in Fig. 6.112 (compare

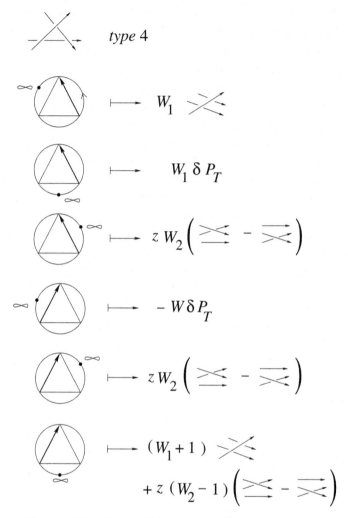

type 4

$$W_1$$

$$W_1\,\delta\,P_T$$

$$z\,W_2\left(\;\;-\;\;\right)$$

$$-\,W\,\delta\,P_T$$

$$z\,W_2\left(\;\;-\;\;\right)$$

$$(W_1+1)\;$$

$$+\,z\,(W_2-1)\left(\;\;-\;\;\right)$$

Figure 6.106:   The partial smoothings for R III of type 4.

Ref. [45]) and the calculation of $\bar{R}^{(1)}$ is shown in Fig. 6.113 (compare Remark 6.10).

We calculate $\bar{R}^{(1)}(rotK) = \delta P_K$ for K the figure eight knot by using the alternative loop from Remark 6.11. The relevant part is contained in Fig. 6.114 and the calculation of $V^{(1)}$ from the relevant part of the Gauss diagrams is given in Fig. 6.115.

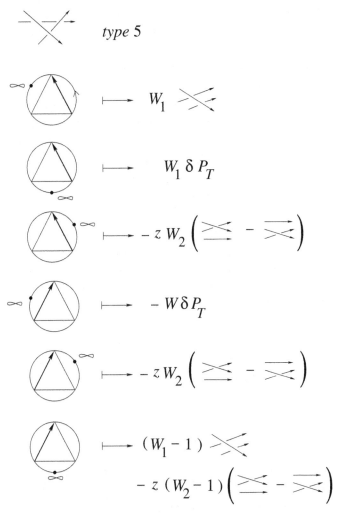

type 5

Figure 6.107: The partial smoothings for R III of type 5.

Remark 6.10 *The scan-arc* $scan(K)$ *is the first half of* $rot(K)$. *In the second half, we always move from* ∞ *to the overcross of a distinguished crossing* d. *Consequently, strata of type* l_c *and of type* r_a *do not occur. Moreover, from* ∞ *we move only over the rest of the diagram up to the overcross of* d. *Hence, there are no f-crossings at all for the strata of type* l_b, r_b *and* II_0. *It follows that the second half does not contribute to* $R^{(1)}(rot(K))$. *There aren't any strata of*

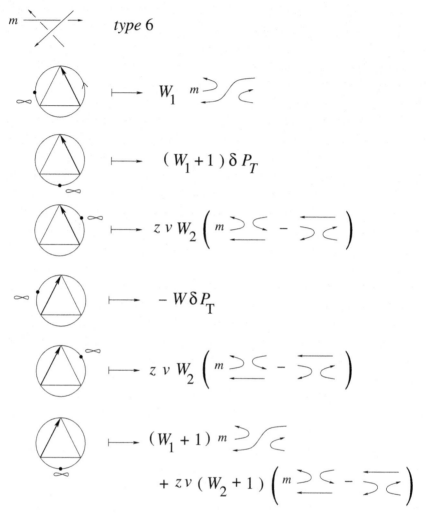

Figure 6.108: The partial smoothings for R III of type 6.

type r_c, l_a, l_c, II_1 *at all in* $rot(K)$ *and the contributions of the two Reidemeister I moves cancel out. Consequently,* $V^{(1)}$ *doesn't contribute and we obtain*

$$\bar{R}^{(1)}(rot(K)) = R^{(1)}(\text{scan}(K))$$

for all long knots K.

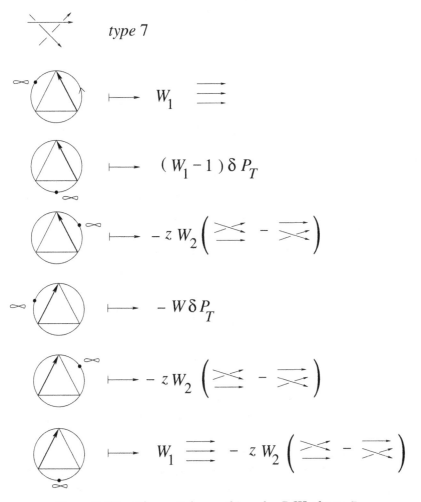

Figure 6.109: The partial smoothings for R III of type 7.

For a general tangle T, we could consider the scan-arc for all components simultaneously as shown in Fig. 6.116. This scan-arc generalises $rot(K)$ for long knots. Let us call it $rot(T)$. Does $R^{(1)}(rot(T))$ also contain $P_T \in H_n(z, v)$ as a factor for each choice of a point at infinity in ∂T and each abstract closure σ of T to a circle?

Example 6.4 The loop drag 3_1^+ is a regular isotopy. Note that many Reidemeister moves do not contribute to $R^{(1)}$ because they do not

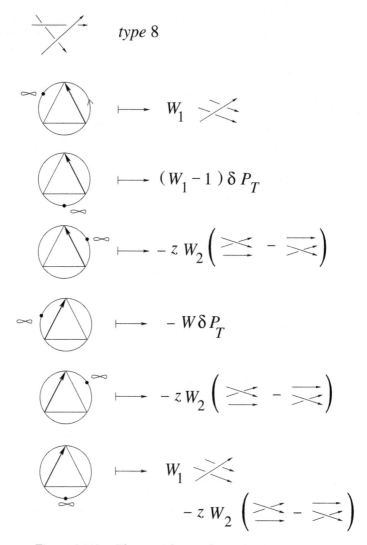

Figure 6.110:   The partial smoothings for R III of type 8.

have the right global type or they have the weight zero but they contribute to $V^{(1)}$. We leave the calculation of $\bar{R}^{(1)}(\text{drag } 3_1^+) = -3\delta P_{(3_1^+)}^2$ to the reader as an exercise.

Remark 6.11 *Let us consider a loop which represents* $rot(K)$ *but by using a positive curl with positive Whitney index as shown*

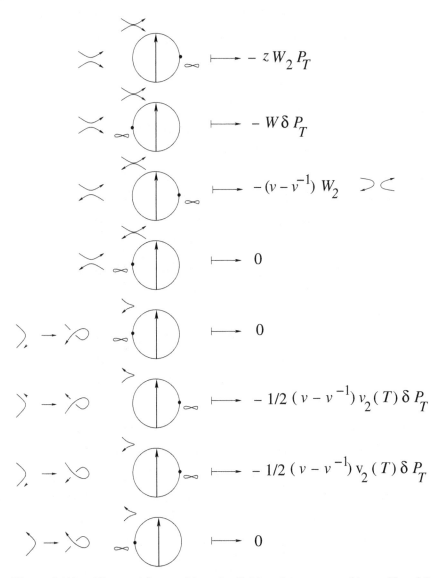

Figure 6.111: The partial smoothings for Reidemeister moves of types II and I.

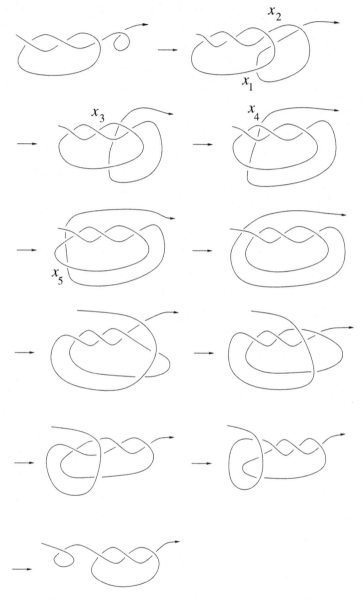

Figure 6.112:   The regular part of the loop $rot(3_1^+)$.

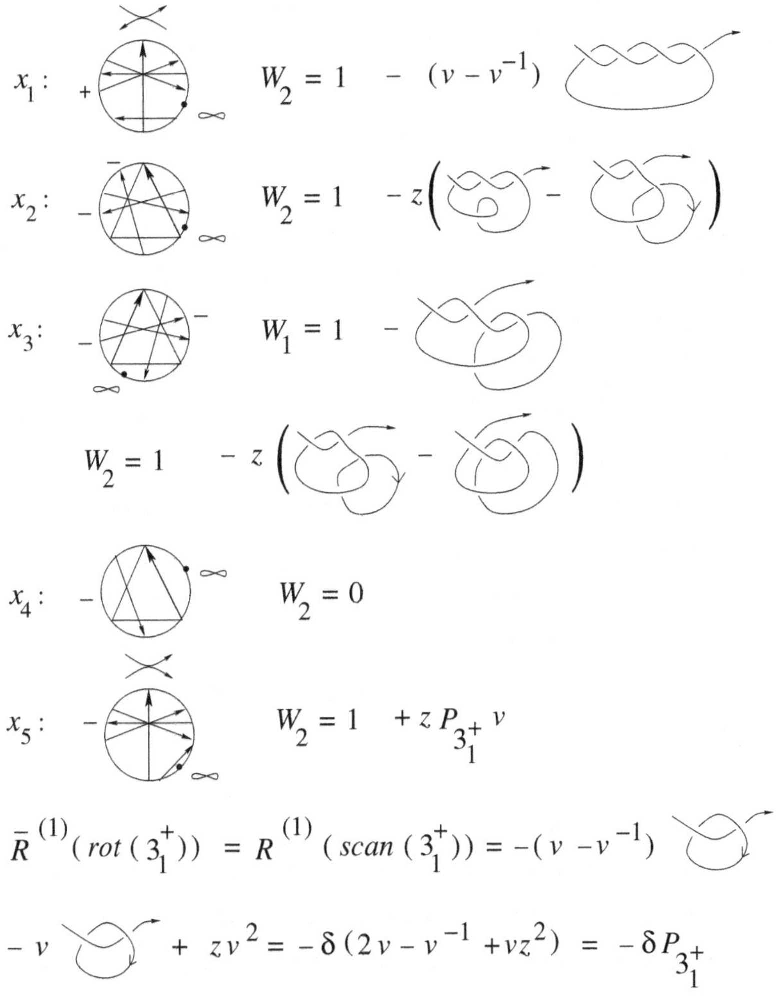

$$x_1: \quad + \qquad W_2 = 1 \quad - \quad (v - v^{-1})$$

$$x_2: \quad - \qquad W_2 = 1 \quad - z \left(\quad - \quad \right)$$

$$x_3: \quad - \qquad W_1 = 1 \quad -$$

$$W_2 = 1 \quad - z \left(\quad - \quad \right)$$

$$x_4: \quad - \qquad W_2 = 0$$

$$x_5: \quad - \qquad W_2 = 1 \quad + z\, P_{3_1^+}\, v$$

$$\bar{R}^{(1)}(\,rot\,(3_1^+)\,) \;=\; R^{(1)}(\,scan\,(3_1^+)\,) = -(v - v^{-1})$$

$$- v \qquad + \; z v^2 = -\delta\,(2v - v^{-1} + v z^2) \;=\; -\delta\, P_{3_1^+}$$

Figure 6.113:   Calculation of $\bar{R}^{(1)}(\mathrm{scan}(3_1^+))$.

in Fig. 6.117. We see immediately that no Reidemeister move at all contributes to $R^{(1)}$. Hence, $\bar{R}^{(1)}(rotK) = -V^{(1)}(rot(K))\delta P_K$. This means that Conjecture 6.1 is not really difficult. It reduces to show that

$$V^{(1)}(rot(K)) = v_2(K)$$

for this loop representing $rot(K)$.

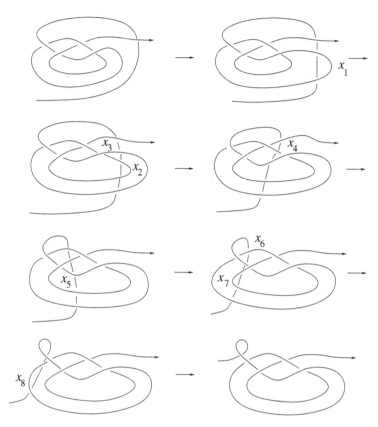

Figure 6.114:   The second half of an alternative loop for $rot(4_1)$.

6.12   Solutions of the Positive Global Tetrahedron Equation for the Kauffman Polynomial

The types, the signs, the weights, and the grading in the Kauffman case are exactly the same as in the HOMFLYPT case. Only the partial smoothings are different.

Proposition 6.8 *The 1-cochain*

$$R_{F,\mathrm{reg}}(m)(A) = \sum_{p \in m} \mathrm{sign}(p)\sigma_2\sigma_1(p)$$

$$+ \sum_{p \in m} \mathrm{sign}(p)W(p)(\sigma_2^2(p) - \sigma_1^2(p))$$

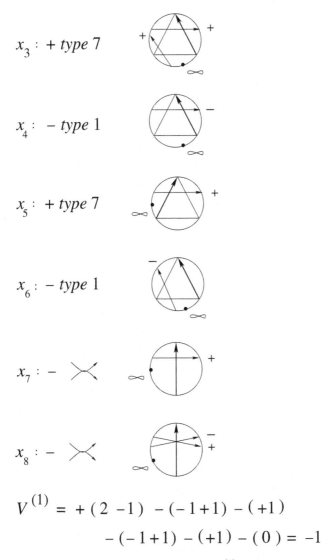

x_3 : + *type* 7

x_4 : − *type* 1

x_5 : + *type* 7

x_6 : − *type* 1

x_7 : −

x_8 : −

$$V^{(1)} = + (2\ {-}1) - (-1{+}1) - (\ {+}1\ )$$
$$- (-1{+}1) - ({+}1) - (\ 0\ ) = -1$$

Figure 6.115: Calculation of $V^{(1)}(rot(4_1))$.

is a solution of the positive global tetrahedron equation. Here, the first sum is over all triple crossings of the global types shown in Fig. 6.15 (*i.e.*, the types l_c and r_a) and such that $\partial(hm) = A$. The second sum is over all triple crossings p which have a distinguished crossing d of type 0 (*i.e.*, the types r_a, r_b and l_b).

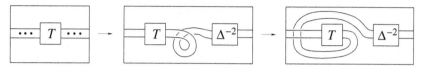

Figure 6.116:   The scan-arc $rot(T)$.

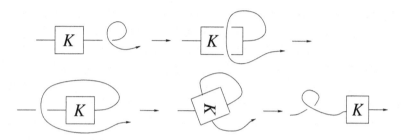

Figure 6.117:   A different loop representing $rot(K)$.

$$t_1 = \quad \quad \quad \quad \quad t_2 = \quad \quad \quad$$

Figure 6.118:   New generators of the Kauffman skein module of 3-braids.

Lemma 6.16 $(\sigma_2^2(p) - \sigma_1^2(p))$ *is a solution with constant weight of the positive tetrahedron equation.*

Proof.   Let t_2 and t_1 be the new generators of the Kauffman skein module of 3-braids as shown in Fig. 6.118. Using the Kauffman skein relations, we see that $(\sigma_2^2(p) - \sigma_1^2(p))$ is equal to $z(\sigma_2(p) - \sigma_1(p)) + zv^{-1}(t_2(p) - t_1(p))$. Like in the HOMFLYPT case, the strata $P_2, \bar{P}_2, P_3, \bar{P}_3$ do not contribute to the solution with constant weight. We give the calculation of $(\sigma_2(p) - \sigma_1(p))$ in Figs. 6.119 and 6.120 and the calculation of $(t_2(p) - t_1(p))$ in Figs. 6.121 and 6.122. We see that they cancel out.   □

Lemma 6.17 $\sigma_2\sigma_1(p)$ *cancels out in* $P_4 - \bar{P}_4$.

Proof.   The proof in the HOMFLYPT case doesn't use the skein relations and hence it is still valid in the Kauffman case.   □

The calculation of $\sigma_2\sigma_1(p)$ for $-P_1 + \bar{P}_1$ is given in Fig. 6.123.

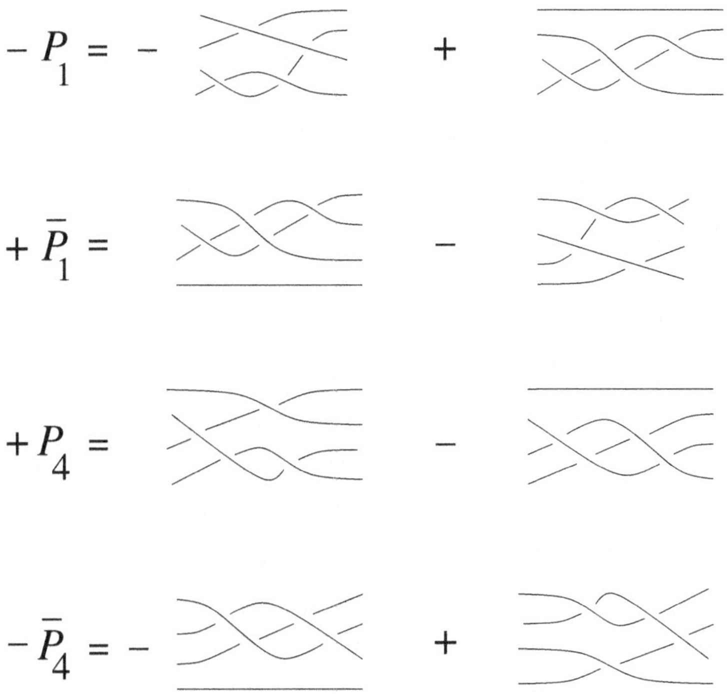

$$- P_1 = - \quad \text{[diagram]} \quad + \quad \text{[diagram]}$$

$$+ \bar{P}_1 = \quad \text{[diagram]} \quad - \quad \text{[diagram]}$$

$$+ P_4 = \quad \text{[diagram]} \quad - \quad \text{[diagram]}$$

$$- \bar{P}_4 = - \quad \text{[diagram]} \quad + \quad \text{[diagram]}$$

Figure 6.119:   Contributions of $\sigma_2(p) - \sigma_1(p)$.

Proof of Proposition 6.8. It is completely analogous to the proof of Proposition 6.2 besides the calculations for $-P_1 + \bar{P}_1 + P_3 - \bar{P}_3$. We give the remaining calculation of the contribution of $(\sigma_2(p) - \sigma_1(p)) + v^{-1}(t_2(p) - t_1(p))$ from P_3 in Fig. 6.124.   □

We haven't solved the cube equations in this case. It seems to us that one should perhaps associate to a self-tangency with equal tangent direction (i.e., of type II_0^-) a partial smoothing which consists of some combination of the four 2-tangles involved in the Kauffman skein relation. But we haven't carried this out.

Proposition 6.9 *The 1-cochain*

$$R_F(m) = \sum_{p \in m} \operatorname{sign}(p) W_1(p) \sigma_2 \sigma_1(p)$$

$$+ \sum_{p \in m} \operatorname{sign}(p) W_2(p) (\sigma_2^2(p) - \sigma_1^2(p))$$

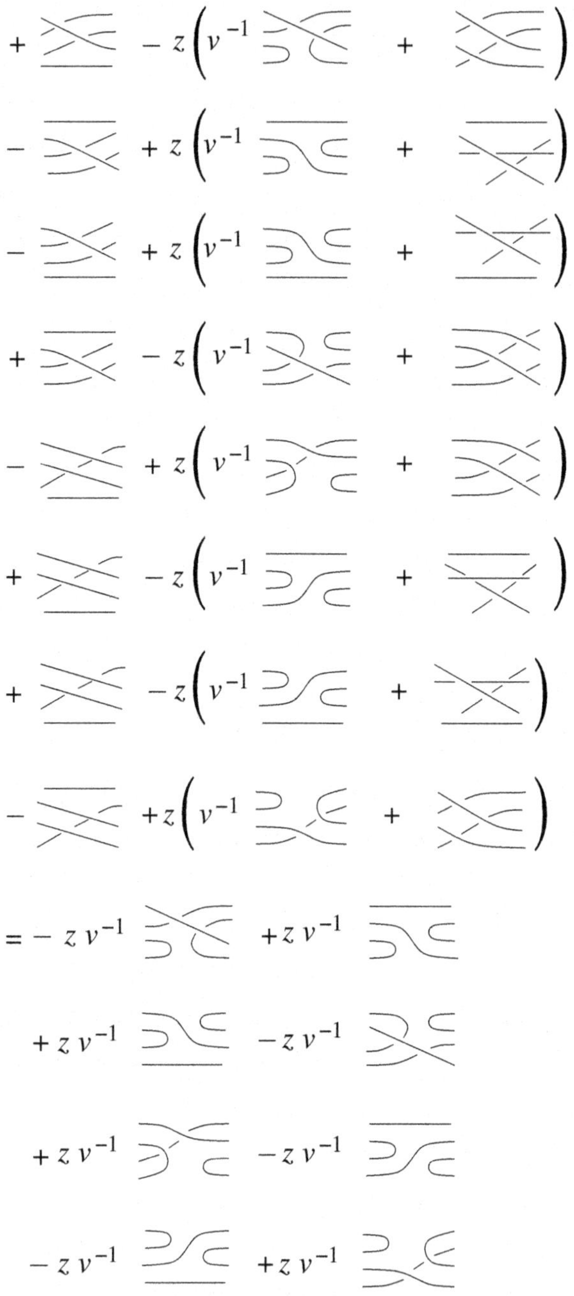

$$+ \quad - z\left(v^{-1} \quad + \quad \right)$$

$$- \quad + z\left(v^{-1} \quad + \quad \right)$$

$$- \quad + z\left(v^{-1} \quad + \quad \right)$$

$$+ \quad - z\left(v^{-1} \quad + \quad \right)$$

$$- \quad + z\left(v^{-1} \quad + \quad \right)$$

$$+ \quad - z\left(v^{-1} \quad + \quad \right)$$

$$+ \quad - z\left(v^{-1} \quad + \quad \right)$$

$$- \quad + z\left(v^{-1} \quad + \quad \right)$$

$$= -z\,v^{-1} \quad + z\,v^{-1} \quad$$

$$+ z\,v^{-1} \quad - z\,v^{-1} \quad$$

$$+ z\,v^{-1} \quad - z\,v^{-1} \quad$$

$$- z\,v^{-1} \quad + z\,v^{-1} \quad$$

Figure 6.120: Calculation of $\sigma_2(p) - \sigma_1(p)$ for $-P_1 + \bar{P}_1 + P_4 - \bar{P}_4$.

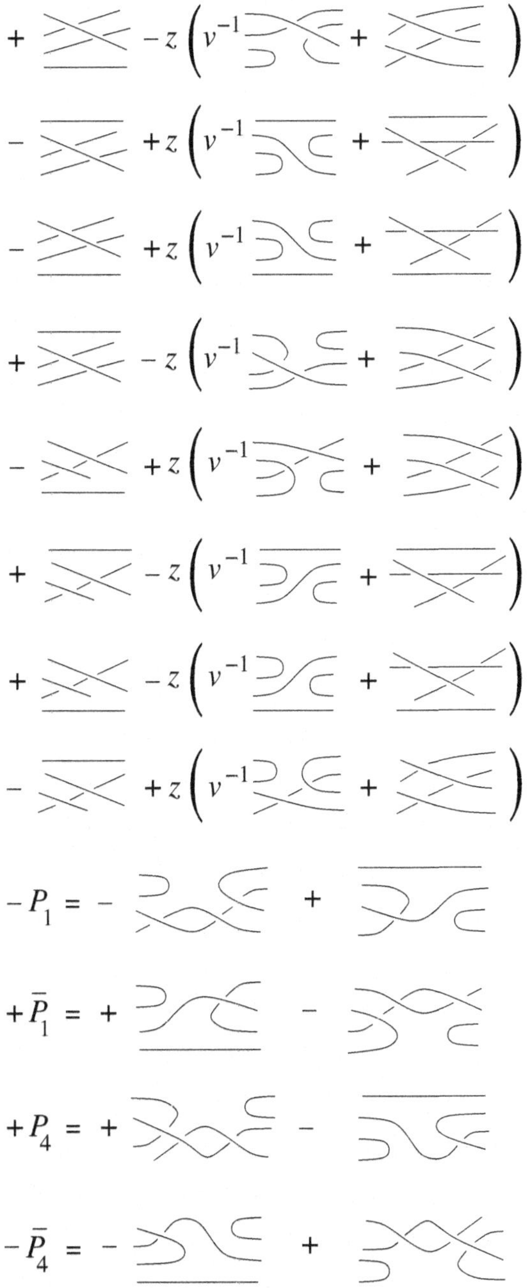

Figure 6.121: Contributions of $t_2(p) - t_1(p)$.

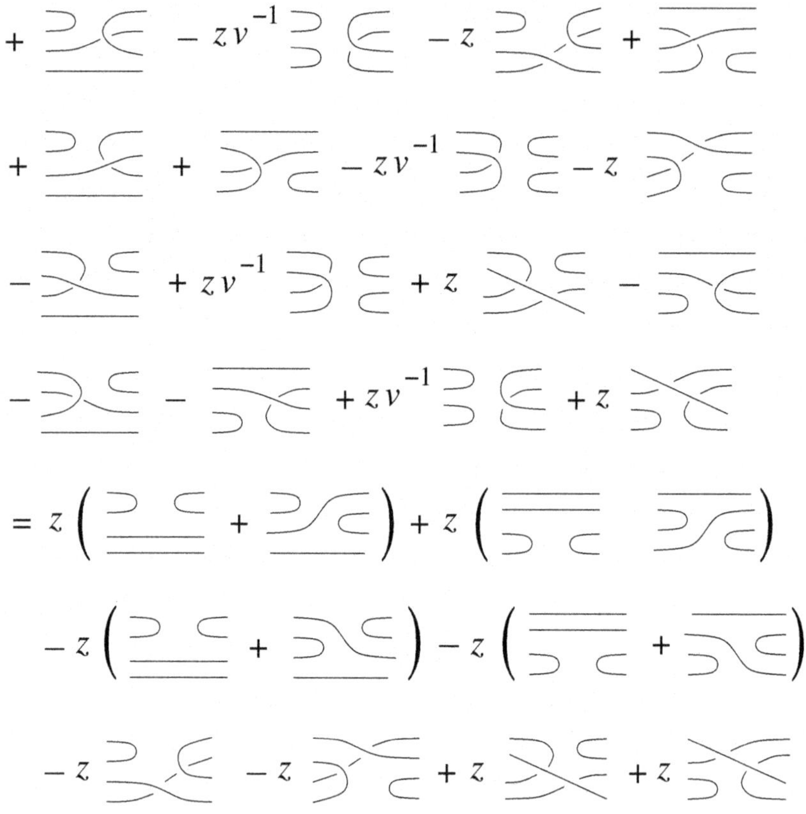

$+ \;\;\cdots\;\; -zv^{-1}\;\;\cdots\;\; -z\;\;\cdots\;\; +\;\;\cdots$

$+ \;\;\cdots\;\; +\;\;\cdots\;\; -zv^{-1}\;\;\cdots\;\; -z\;\;\cdots$

$- \;\;\cdots\;\; +zv^{-1}\;\;\cdots\;\; +z\;\;\cdots\;\; -\;\;\cdots$

$- \;\;\cdots\;\; -\;\;\cdots\;\; +zv^{-1}\;\;\cdots\;\; +z\;\;\cdots$

$= z\left(\cdots + \cdots\right) + z\left(\cdots\quad\cdots\right)$

$- z\left(\cdots + \cdots\right) - z\left(\cdots + \cdots\right)$

$- z\;\;\cdots\;\; -z\;\;\cdots\;\; +z\;\;\cdots\;\; +z\;\;\cdots$

Figure 6.122: Calculation of $t_2(p) - t_1(p)$ for $-P_1 + \bar P_1 + P_4 - \bar P_4$.

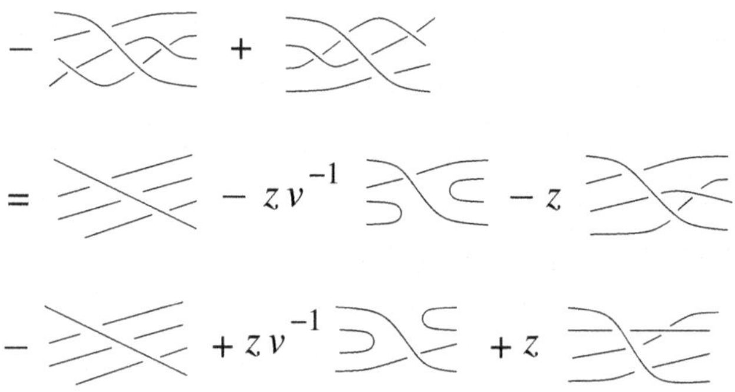

$- \;\;\cdots\;\; +\;\;\cdots$

$= \;\;\cdots\;\; -zv^{-1}\;\;\cdots\;\; -z\;\;\cdots$

$- \;\;\cdots\;\; +zv^{-1}\;\;\cdots\;\; +z\;\;\cdots$

Figure 6.123: Calculation of $\sigma_2\sigma_1(p)$ for $-P_1 + \bar P_1$.

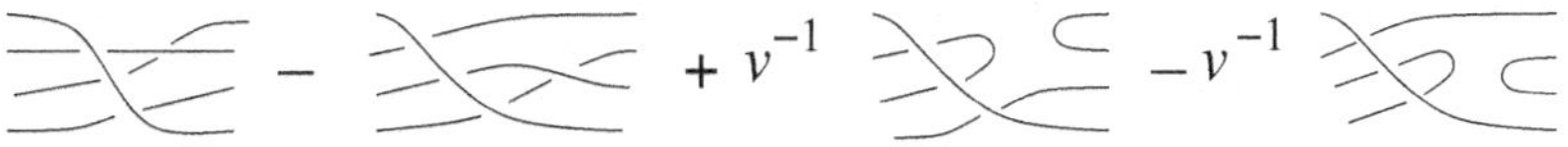

Figure 6.124:   Calculation of $(\sigma_2(p) - \sigma_1(p)) + v^{-1}(t_2(p) - t_1(p))$ for P_3.

is a solution of the positive global tetrahedron equation. Here, the first sum is over all triple crossings of the global types as shown in Fig. 6.15 (i.e., the types l_c and r_a). The second sum is over all triple crossings p which have a distinguished crossing d of type 0 (i.e., the types r_a, r_b and l_b).

Proof.   This follows immediately from the proofs of Propositions 6.8 and 6.3. ☐

Surprisingly, there is a second solution of the global positive tetrahedron equation in Kauffman's case.

Proposition 6.10 *The 1-cochain*

$$R^{(1)}_{F,\text{reg}}(m)(A) = \sum_{p \in m} \text{sign}(p)t_1t_2(p)$$

$$+ \sum_{p \in m} \text{sign}(p)W(p)z(-\sigma_1^{-1}t_2 - \sigma_2 t_1 + t_1\sigma_2^{-1} + t_2\sigma_1)$$

(compare Fig. 6.8, where we give it in shorter form) is also a solution of the positive global tetrahedron equation. Here, the first sum is over all triple crossings of the global types shown in Fig. 6.15 (i.e., the types l_c and r_a) and such that $\partial(hm) = A$. The second sum is over all triple crossings p which have a distinguished crossing d of type 0 (i.e., the types r_a, r_b and l_b).

Lemma 6.18 $(\sigma_1^{-1}t_2 + \sigma_2 t_1 - t_1\sigma_2^{-1} - t_2\sigma_1)$ *is a solution with constant weight of the positive tetrahedron equation.*

Proof.   The calculations are contained in Figs. 6.125–6.127. ☐

Lemma 6.19 $t_1t_2(p)$ *cancels out in $P_4 - \bar{P}_4$.*

Proof.   The proof is in Fig. 6.128. ☐

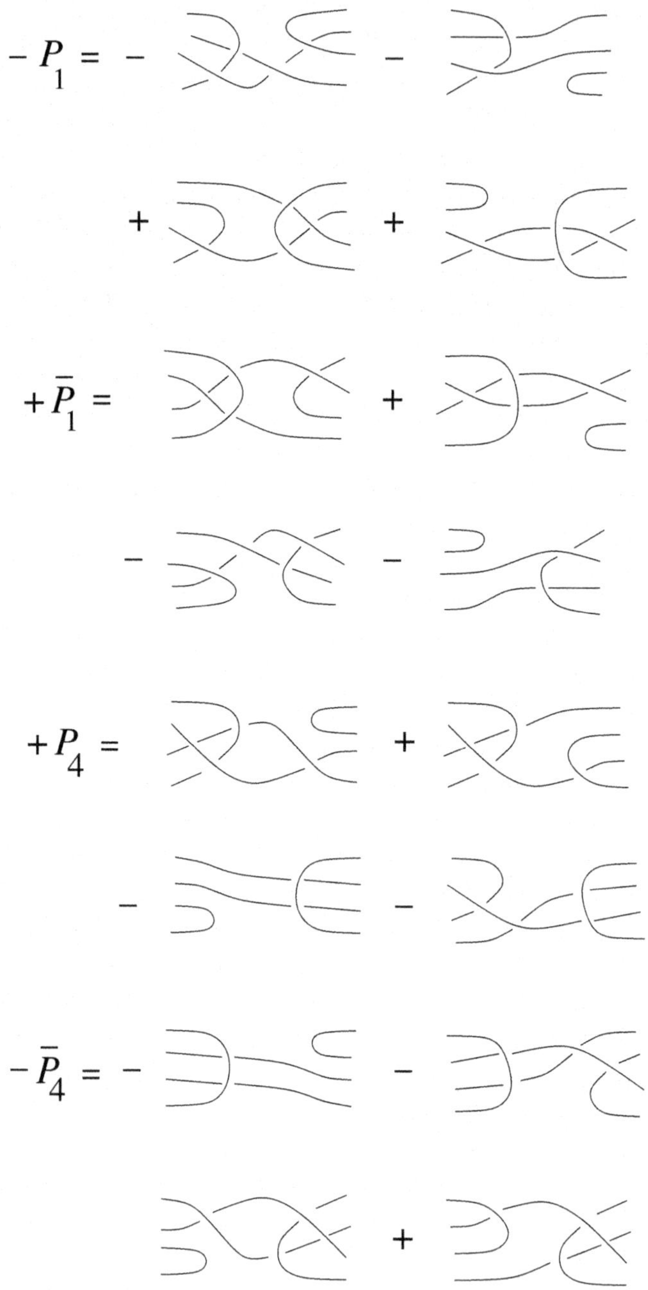

$$-\,P_1 = -\ \ \text{(diagram)}\ -\ \text{(diagram)}$$

$$+\ \text{(diagram)}\ +\ \text{(diagram)}$$

$$+\,\bar{P}_1 =\ \text{(diagram)}\ +\ \text{(diagram)}$$

$$-\ \text{(diagram)}\ -\ \text{(diagram)}$$

$$+\,P_4 =\ \text{(diagram)}\ +\ \text{(diagram)}$$

$$-\ \text{(diagram)}\ -\ \text{(diagram)}$$

$$-\,\bar{P}_4 = -\ \text{(diagram)}\ -\ \text{(diagram)}$$

$$\text{(diagram)}\ +\ \text{(diagram)}$$

Figure 6.125:   Contribution of $\sigma_1^{-1}t_2 + \sigma_2 t_1 - t_1\sigma_2^{-1} - t_2\sigma_1$.

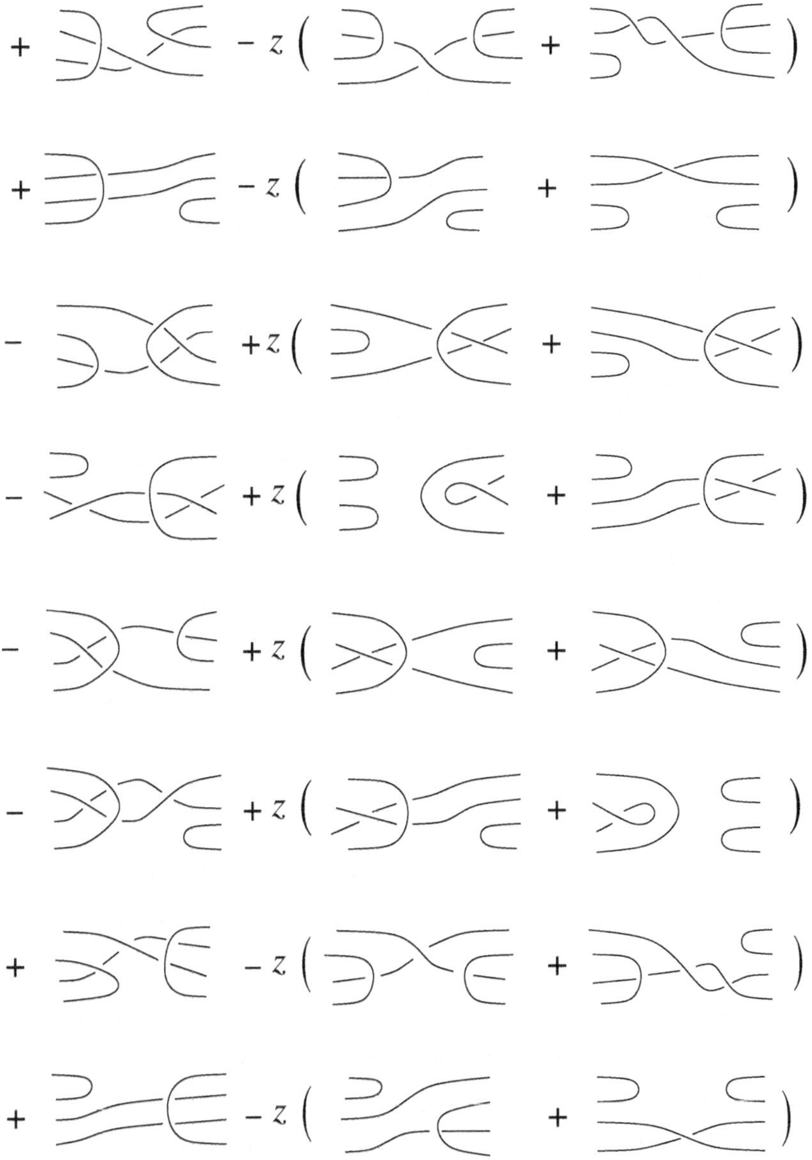

Figure 6.126: Calculation of $\sigma_1^{-1}t_2 + \sigma_2t_1 - t_1\sigma_2^{-1} - t_2\sigma_1$ for $-P_1 + \bar{P}_1$.

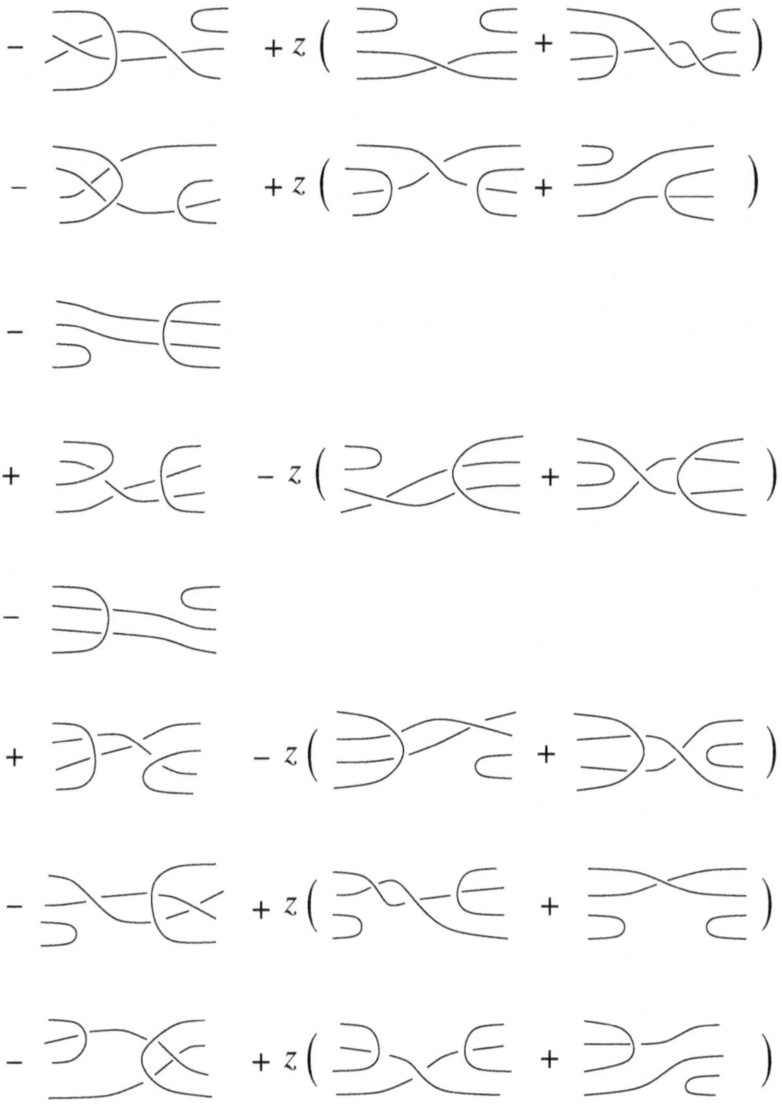

Figure 6.127:   Calculation of $\sigma_1^{-1}t_2 + \sigma_2 t_1 - t_1\sigma_2^{-1} - t_2\sigma_1$ for $P_4 - \bar{P}_4$.

Proof of Proposition 6.10. The contribution of $-P_1 + \bar{P}_1$ is shown in Fig. 6.129 and the remaining contribution of $(\sigma_1^{-1}t_2 + \sigma_2 t_1 - t_1\sigma_2^{-1} - t_2\sigma_1)$ from P_3 is shown in Fig. 6.130. We see that their combination in $R_{F,\mathrm{reg}}^{(1)}(m)(A)$ cancels out. $\qquad\square$

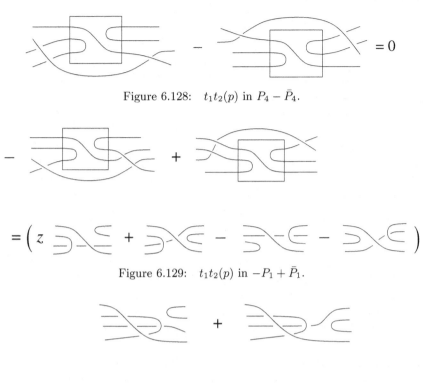

Figure 6.128: $t_1 t_2(p)$ in $P_4 - \bar{P}_4$.

Figure 6.129: $t_1 t_2(p)$ in $-P_1 + \bar{P}_1$.

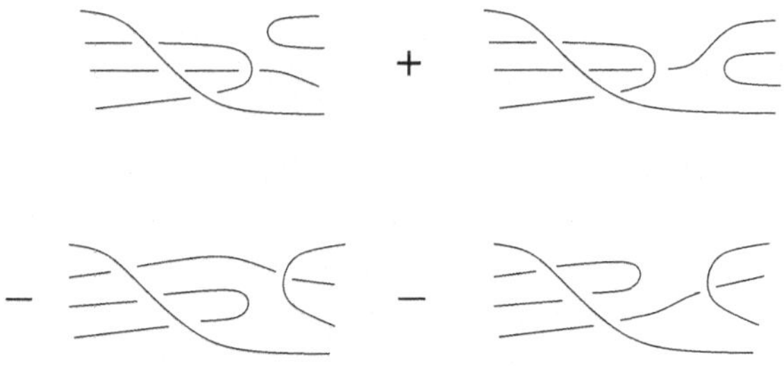

Figure 6.130:   Calculation of $\sigma_1^{-1} t_2 + \sigma_2 t_1 - t_1 \sigma_2^{-1} - t_2 \sigma_1$ in P_3.

Remark 6.12 *The Kauffman polynomial as an invariant of regular isotopy depends only on the unoriented link in contrast to our 1-cocycle $R^{(1)}_{F,\text{reg}}(m)(A)$. Indeed, almost all our definitions use in an essential way the orientation of the circle $T \cup \sigma$. Besides the signs of the Reidemeister moves, they are not even invariant under flip.*

Proposition 6.11 *The 1-cochain*

$$R^{(1)}_F(m) = \sum_{p \in m} \text{sign}(p) W_1(p) t_1 t_2(p)$$

$$+ \sum_{p \in m} \text{sign}(p) W_2(p) z(-\sigma_1^{-1} t_2 - \sigma_2 t_1 + t_1 \sigma_2^{-1} + t_2 \sigma_1)$$

is a solution of the positive global tetrahedron equation. Here, the first sum is over all triple crossings of the global types shown in Fig. 6.15 (i.e., the types l_c and r_a). The second sum is over all triple crossings p which have a distinguished crossing d of type 0 (i.e., the types r_a, r_b and l_b).

Proof. This follows immediately from the proofs of Propositions 6.10 and 6.3. $\qquad\qquad\qquad\qquad\qquad\qquad\qquad\qquad\qquad\qquad\qquad$ $\square$

6.13   The 1-Cocycles $R^{(1)}_{F,\mathrm{reg}}$, $R^{(1)}_F$ and $\bar{R}^{(1)}_F$

We proceed in a way completely analogous to the case of the HOM-FLYPT invariant. However, we will solve the cube equations only with coefficients in $\mathbb{Z}/2\mathbb{Z}$ (again, one has to probably associate more general partial smoothings with self-tangencies in order to get a solution with integer coefficients).

Definition 6.28 *For $R^{(1)}_{F,\mathrm{reg}}$:*
The partial smoothing $T_{II_0^+}(p)$ of a self-tangency with opposite tangent direction and d of type 0 is defined in Fig. 6.131.
The partial smoothing $T_{II_0^-}(p)$ of a self-tangency with equal tangent direction and d of type 0 is defined in Fig. 6.132.

For $R^{(1)}_F$:
We replace in the above definitions $W(p)$ by $W_2(p)$ and we normalise as usual F_T by $v^{-w(T)}$.
The partial smoothing $T_I(p)$ of a Reidemeister I move with d of type 0 is defined in Fig. 6.133.

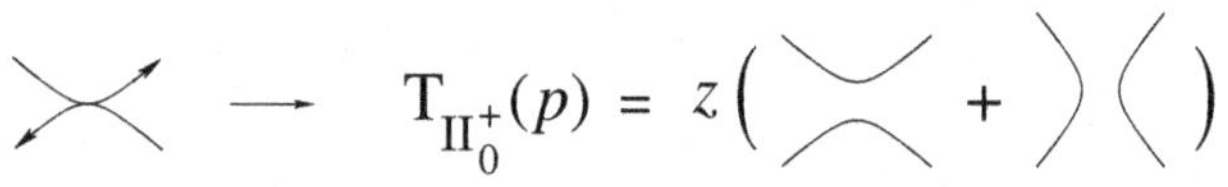

$$\times \longrightarrow T_{II_0^+}(p) = z\left(\smile\frown + \;)(\; \right)$$

Figure 6.131:   Partial smoothing of an R II move with opposite tangent direction.

$$\times \longrightarrow T_{II_0^-}(p) = (v + v^{-1})\;)(\;$$

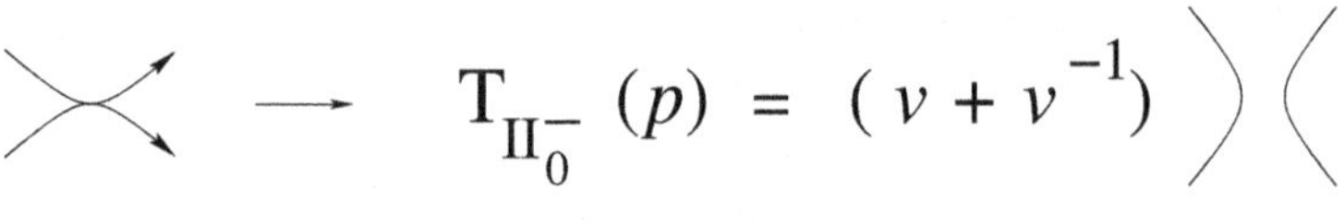

Figure 6.132:   Partial smoothing of an R II move with same tangent direction.

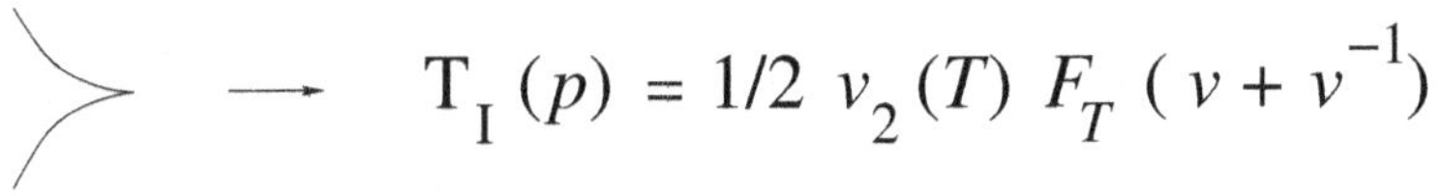

$$T_I(p) = 1/2\, v_2(T)\, F_T\, (v + v^{-1})$$

Figure 6.133: Partial smoothing of an $R\,\mathrm{I}$ move.

Definition 6.29 *Let p be a self-tangency with opposite tangent direction. Its contribution to $R^{(1)}_{F,\mathrm{reg}}$ is defined by*

$$R^{(1)}_{F,\mathrm{reg}} = \mathrm{sign}(p)W(p)T_{II_0^+}(p).$$

Its contribution to $R^{(1)}_F$ is defined by

$$R^{(1)}_F = \mathrm{sign}(p)W_2(p)T_{II_0^+}(p).$$

Let p be a self-tangency with equal tangent direction. Its contribution to $R^{(1)}_{F,\mathrm{reg}}$ is defined by

$$R^{(1)}_{F,\mathrm{reg}} = \mathrm{sign}(p)W(p)T_{II_0^-}(p).$$

Its contribution to $R^{(1)}_F$ is defined by

$$R^{(1)}_F = \mathrm{sign}(p)W_2(p)T_{II_0^-}(p).$$

The contribution of a Reidemeister I *move with d of type 0 is defined by*

$$R^{(1)}_F = \mathrm{sign}(p)T_I(p).$$

Lemma 6.20 *Let m be the meridian of $\Sigma^{(2)}_{\mathrm{self-flex}}$ (compare Section 6.2). Then, $R^{(1)}_{F,\mathrm{reg}}(m) = 0$ and $R^{(1)}_F(m) = 0$.*

Proof. The weights are exactly the same as in the HOMFLYPT case. We calculate the values on the meridians in Fig. 6.134. $\square$

Lemma 6.21 *The value of the 1-cocycle $R^{(1)}_F$ on a meridian of a degenerate cusp, locally given by $x^2 = y^5$ and denoted by $\Sigma^{(2)}_{\mathrm{cusp-deg}}$, is zero.*

Proof. Again, the cusp can be of type 0 or of type 1. Only the case of type 0 is interesting and we give the calculation in Fig. 6.135. $\square$

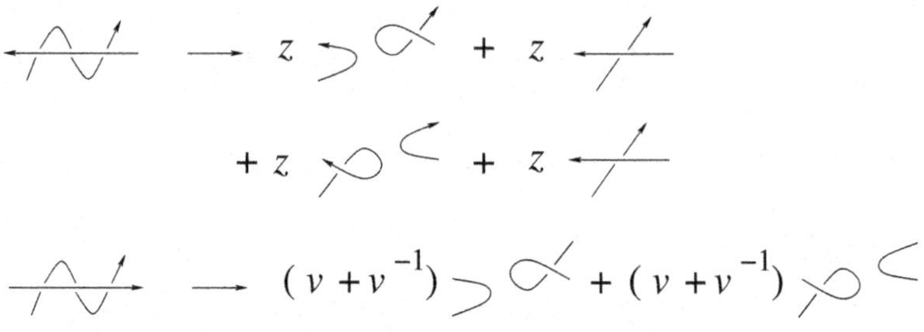

Figure 6.134: The meridians of a self-tangency in a flex.

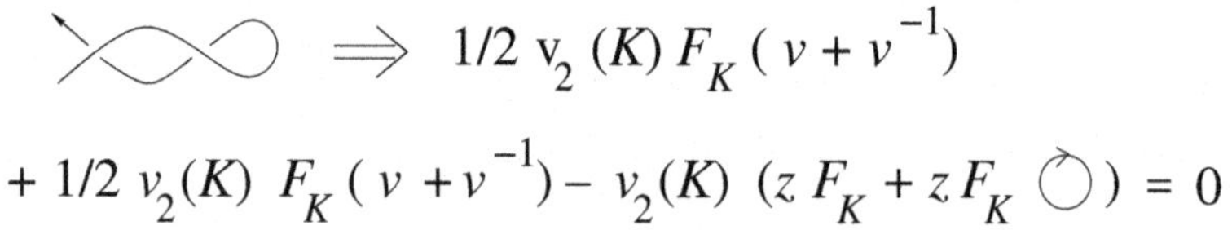

$$\Rightarrow \ 1/2 \ v_2 \, (K) \, F_K \, (\, v + v^{-1})$$

$$+ \ 1/2 \ v_2(K) \ F_K \, (\, v + v^{-1}) - v_2(K) \ (z \, F_K + z \, F_K \bigcirc) = 0$$

Figure 6.135: $R_F^{(1)}$ for the meridians of a degenerated cusp.

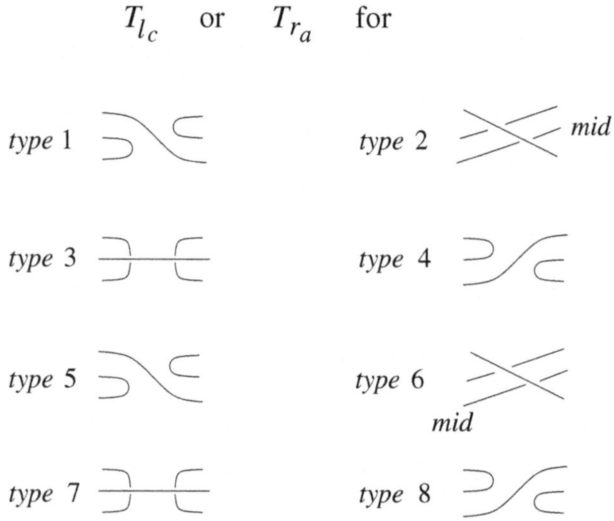

$$T_{l_c} \quad \text{or} \quad T_{r_a} \quad \text{for}$$

Figure 6.136: The partial smoothings T_{l_c} and T_{r_a} in the Kauffman case.

We will now solve the cube equations.

Definition 6.30 *The partial smoothings for the local and global types of triple crossings in Kauffman's case are given in Figs. 6.136 and 6.137 (remember that* mid *denotes the ingoing middle branch for a star-like triple crossing).*

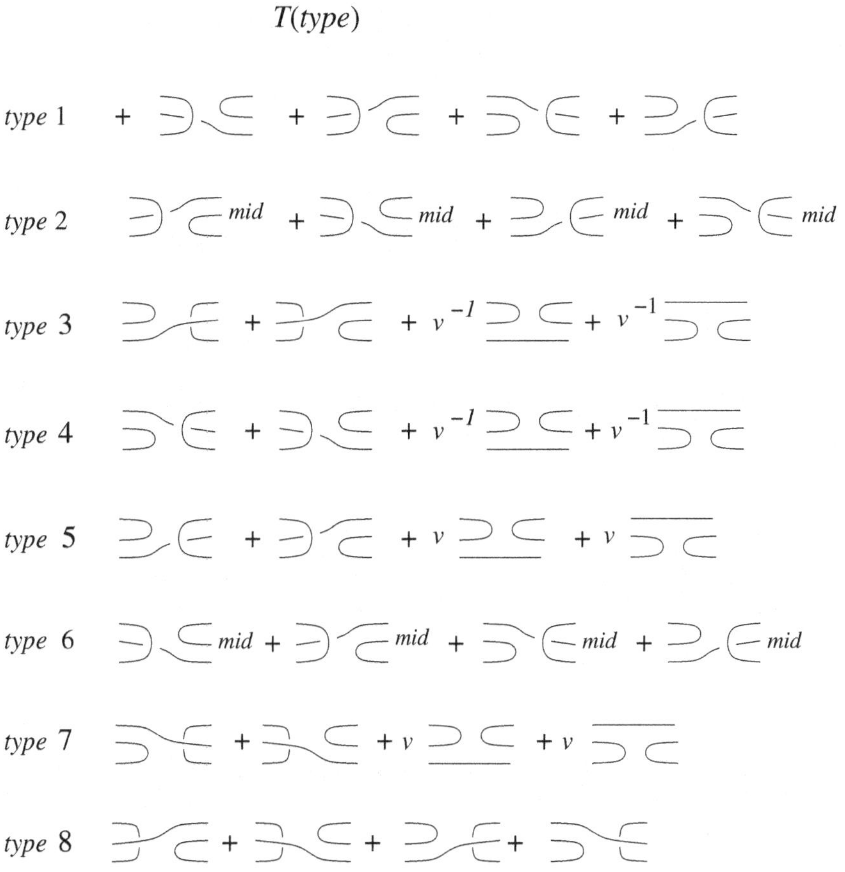

$$T(type)$$

type 1

type 2

type 3

type 4

type 5

type 6

type 7

type 8

Figure 6.137:   The partial smoothings $T(type)$ in the Kauffman case.

Definition 6.31 *Let s be a generic oriented arc in M_T^{reg} (with a fixed abstract closure $T \cup \sigma$ to an oriented circle). Let $A \subset \partial T$ be a given grading. The 1-cochain $R_{F,\mathrm{reg}}^{(1)}(A)$ is defined by*

$$R_{F,\mathrm{reg}}^{(1)}(A)(s) = \sum_{p \in s \cap l_c} \mathrm{sign}(p) T_{l_c}(\text{type})(p)$$

$$+ \sum_{p \in s \cap r_a} \mathrm{sign}(p) T_{r_a}(\text{type})(p)$$

$$+ \sum_{p \in s \cap r_a} \mathrm{sign}(p) z W(p) T(\text{type})(p)$$

$$+ \sum_{p \in s \cap r_b} \text{sign}(p) z W(p) T(\text{type})(p)$$

$$+ \sum_{p \in s \cap l_b} \text{sign}(p) z W(p) T(\text{type})(p)$$

$$+ \sum_{p \in s \cap II_0^+} \text{sign}(p) W(p) T_{II_0^+}(p)$$

$$+ \sum_{p \in s \cap II_0^-} \text{sign}(p) W(p) T_{II_0^-}(p).$$

Here, all weights $W(p)$ are defined only over the f-crossings f with $\partial f = A$ and in the first two sums (i.e., for T_{l_c} and T_{r_a}), we require that $\partial(hm) = A$ for the triple crossings.

Definition 6.32 *Let s be a generic oriented arc in M_T^{reg} (with a fixed abstract closure $T \cup \sigma$ to an oriented circle). The 1-cochain $R_F^{(1)}$ is defined by*

$$R_F^{(1)}(s) = \sum_{p \in s \cap l_c} \text{sign}(p) W_1(p) T_{l_c}(\text{type})(p)$$

$$+ \sum_{p \in s \cap r_a} \text{sign}(p) [W_1(\text{type})(p) T_{r_a}(\text{type})(p)$$

$$+ z W_2(\text{type})(p) T(\text{type})(p)]$$

$$+ \sum_{p \in s \cap r_b} \text{sign}(p) z W_2(p) T(\text{type})(p)$$

$$+ \sum_{p \in s \cap l_b} \text{sign}(p) z W_2(p) T(\text{type})(p)$$

$$+ \sum_{p \in s \cap II_0^+} \text{sign}(p) W_2(p) T_{II_0^+}(p)$$

$$+ \sum_{p \in s \cap II_0^-} \text{sign}(p) W_2(p) T_{II_0^-}(p)$$

$$+ \sum_{p \in s \cap I} \text{sign}(p) T_I(p).$$

Proposition 6.12 *The 1-cochains* $R^{(1)}_{F,\text{reg}}(A)$ *and* $R^{(1)}_F$ *with the adjustments from Definition 6.20 (in Section 6.7) satisfy the cube equations.*

Proof. The weights were already studied for the corresponding proofs in the case of the HOMFLYPT invariant. We just have to check that our partial smoothings satisfy the cube equations. This is completely analogous to the HOMFLYPT case and we left the verification to the reader (using the corresponding figures in Section 6.6). But remember that we consider only coefficients in $\mathbb{Z}/2\mathbb{Z}$. $\qquad \square$

Exactly the same arguments as in the HOMFLYPT case imply that $R^{(1)}_{F,\text{reg}}(A)$ and $R^{(1)}_F$ vanish on the meridians of $\Sigma^{(1)} \cap \Sigma^{(1)}$.

Lemma 6.22 *The value of the 1-cocycle* $R^{(1)}_F$ *on a meridian of* $\Sigma^{(2)}_{\text{trans}-\text{cusp}}$ *is zero besides for* $\Sigma^{(2)}_{l_c}$.

Proof. We consider just one case shown in Fig. 6.138. All other cases are similar and we left the verification to the reader. $\qquad \square$

$R^{(1)}_{F,\text{reg}}(A)$ and $R^{(1)}_F$ have both the scan-property for branches moving under the tangle T. The proof is exactly the same as in the HOMFLYPT case. Consequently, we have proven the following proposition.

Proposition 6.13 *The 1-cochains* $R^{(1)}_{F,\text{reg}}(A)$ *and* $R^{(1)}_F$ *are 1-cocycles in* M^{reg}_T, *respectively,* $M_T \backslash \Sigma^{(2)}_{l_c}$. *They both have the scan-property.*

Definition 6.33 *The completion* $\bar{R}^{(1)}_F$ *mod 2 is defined by* $\bar{R}^{(1)}_F = R^{(1)}_F + F_T V^{(1)}$. *Here,* $V^{(1)}$ *is the same as in the HOMFLYPT case.*

Theorem 6.4 $\bar{R}^{(1)}_F$ *is a 1-cocycle in* M_T *which represents a nontrivial cohomology class and which has the scan-property for branches moving under the tangle* T.

Proof. The value of $R^{(1)}_F$ on the meridians of $\Sigma^{(2)}_{l_c}$ is equal to $W_1(p)F_T$ as shown in Fig. 6.139. Consequently, $R^{(1)}_F + F_T V^{(1)}$ is already a 1-cocycle for the whole M_T, exactly like in the HOMFLYPT case. $\bar{R}^{(1)}_F$ has the scan-property because both $R^{(1)}_F$ and $V^{(1)}$ have this property. $\qquad \square$

We consider just the easiest example in order to show that $\bar{R}^{(1)}_F$ represents a non-trivial cohomology class.

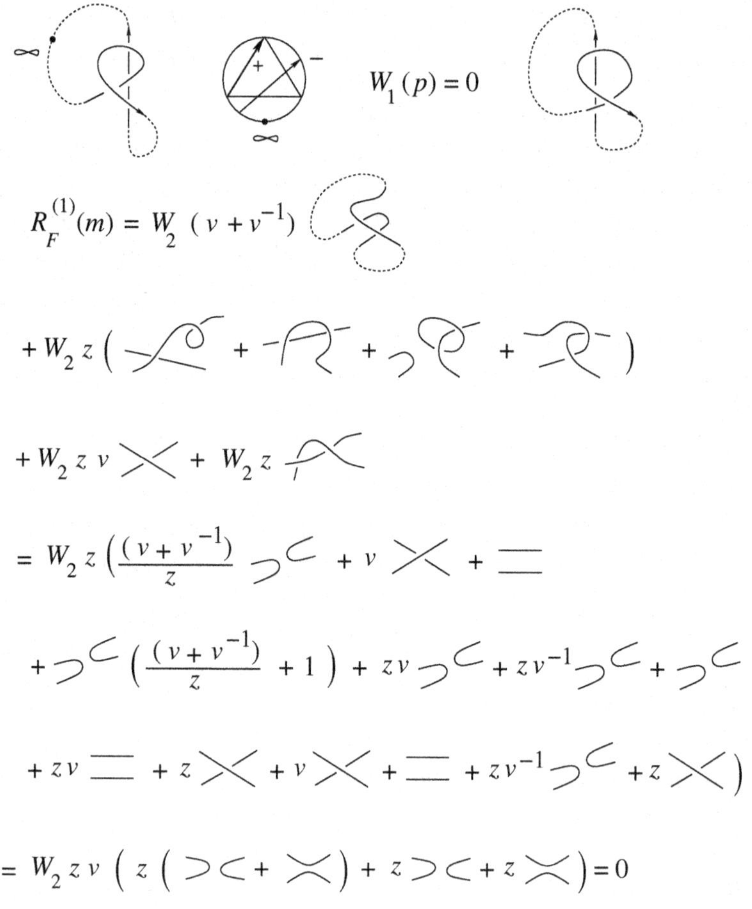

$$W_1(p) = 0$$

$$R_F^{(1)}(m) = W_2(v + v^{-1})$$

$$+ W_2 z \left(+ + + \right)$$

$$+ W_2 z v + W_2 z $$

$$= W_2 z \left(\frac{(v + v^{-1})}{z} + v + \right.$$

$$+ \left(\frac{(v + v^{-1})}{z} + 1 \right) + z v + z v^{-1} + $$

$$\left. + z v + z + v + + z v^{-1} + z \right)$$

$$= W_2 z v \left(z \left(+ \right) + z + z \right) = 0$$

Figure 6.138: $R_F^{(1)}$ on a meridian of a cusp with a transverse branch.

Example 6.5 $\bar{R}_F^{(1)}(rot(3_1^+)) = F_{3_1^+}$. We only have to calculate $\bar{R}_F^{(1)}(\text{scan}(3_1^+))$ because the second half of the rotation does not contribute at all like in the HOMFLYPT case. Moreover, the types and the weights are exactly the same as in the HOMFLYPT case. Only the partial smoothings are different. We give the calculation in Fig. 6.140.

We end with the following interesting question.

Question 6.2 *Do the cube equations have a solution with integer coefficients in Kauffman's case?*

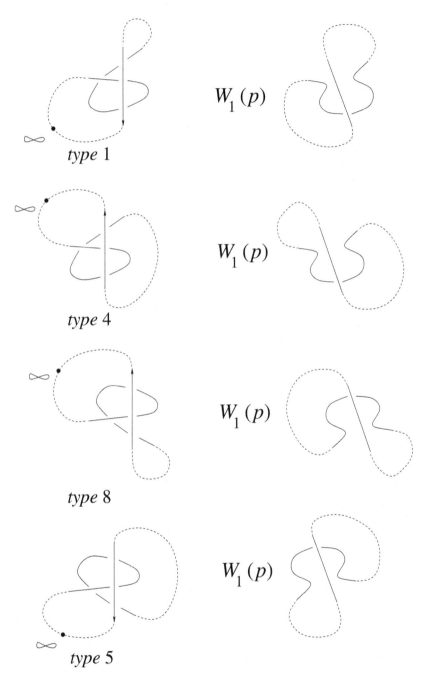

$W_1(p)$

type 1

$W_1(p)$

type 4

$W_1(p)$

type 8

$W_1(p)$

type 5

Figure 6.139: $R_F^{(1)}$ on the meridians of $\Sigma_{l_c}^{(2)}$.

$$+ z \left(F_{3_1^+} v + \begin{array}{c}\includegraphics{}\end{array} \right)$$

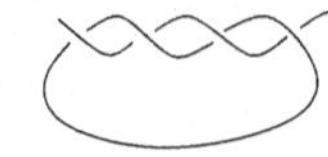

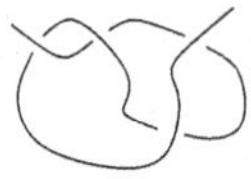

$$+ z \left(\begin{array}{c}\includegraphics{}\end{array} + \begin{array}{c}\includegraphics{}\end{array} \right.$$

$$\left. + \begin{array}{c}\includegraphics{}\end{array} + \begin{array}{c}\includegraphics{}\end{array} \right)$$

$$+ \begin{array}{c}\includegraphics{}\end{array}$$

$$+ z \left(\begin{array}{c}\includegraphics{}\end{array} + \begin{array}{c}\includegraphics{}\end{array} \right.$$

$$\left. + \begin{array}{c}\includegraphics{}\end{array} + \begin{array}{c}\includegraphics{}\end{array} \right)$$

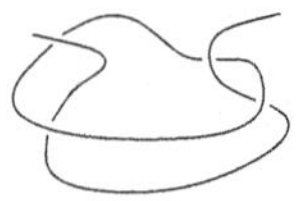

$$+ (v + v^{-1}) \begin{array}{c}\includegraphics{}\end{array}$$

$$= zF_{3_1^+} v + z v^{-2} + z F_{3_1^+} v$$

$$+ z v^{-2} \delta_F + v^{-1} + z \begin{array}{c}\includegraphics{}\end{array}$$

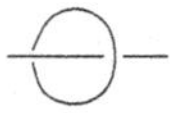

$$+ z v^{-2} + z \delta_F + z + (v + v^{-1}) v^{-2}$$

$$= v^{-1} + z (v^{-2} + 1) + z^2 (v + v^{-1}) = F_{3_1^+} \ mod \ 2$$

Figure 6.140:   Calculation of $R_F^{(1)}(rot(3_1^+))$.

Bibliography

[1] Birman J.: *Braids, Links and Mapping Class Groups*, Annals of Mathematics Studies 82, Princeton University Press, Princeton (1974).

[2] Bar-Natan D.: On the Vassiliev knot invariants, *Topology* 34 (1995), 423–472, Elsevier Science Ltd.

[3] Bar-Natan D., van der Veen R.: A polynomial time knot polynomial, *Proc. Amer. Math. Soc.* 147 (2019), 377–397.

[4] Brandenbursky M.: Link invariants via counting surfaces, *Geom. Dedicata* 173 (2014), 243–270.

[5] Budney R., Conant J., Scannell K., Sinha D.: New perspectives on self-linking, *Adv. Math.* 191 (2005), 78–113.

[6] Budney R., Cohen F.: On the homology of the space of knots, *Geom. Topol.* 13 (2009), 99–139.

[7] Budney R.: Topology of spaces of knots in dimension 3, *Proc. London Math. Soc.* 101 (2010), 477–496.

[8] Budney R.: An operad for splicing, *J. Topol.* 5 (2012), 945–976.

[9] Burde G., Zieschang H.: Knots, de Gruyter Studies in Mathematics 5, Walter de Gruyter GmbH, Berlin, New York (1985).

[10] Carter J.S., Saito M.: Reidemeister moves for surface isotopies and their interpretations as moves to movies, *J. Knot Theory Ramif.* 2 (1993), 251–284.

[11] Chmutov S., Duzhin S., Mostovoy J.: *Introduction to Vassiliev Knot Invariants*, Cambridge University Press, Cambridge (2012).

[12] Chmutov S., Polyak M.: Elementary combinatorics of the HOMFLYPT polynomial, *Inter. Math. Res. Notices* 3 (2010), 480–495.

[13] Chmutov S., Khoury M., Rossi A.: Polyak-Viro formulas for coefficients of the Conway polynomial, *J. Knot Theory Ramif.* 18 (2009), 773–783.

[14] Duzhin S., Karev M.: Detecting the orientation of string links by finite type invariants, *Funct. Anal. Appl.* 41 (2007), 208–216.

[15] Fiedler T.: *Gauss Diagram Invariants for Knots and Links*, Mathematics and Its Applications 532, Kluwer Academic Publishers (2001).

[16] Fiedler T.: *Polynomial One-cocycles for Knots and Closed Braids*, Series on Knots and Everything 64, World Scientific (2019).

[17] Fiedler T.: More 1-cocycles for classical knots, *J. Knot Theory Ramif.* 30 (2021), 57, World Scientific, Singapore.

[18] Fiedler T.: Quantum one-cocycles for knots, arXiv: 1304.0970v2, 177.

[19] Fiedler T., Kurlin V.: A one-parameter approach to links in a solid torus, *J. Math. Soc. Japan* 62 (2010), 167–211.

[20] Fox R.: Rolling, *Bull. Amer. Math. Soc.* 72 (1966), 162–164.

[21] Gordon C., Luecke J.: Knots are determined by their complements. *Bull. Amer. Math. Soc.* 20 (1989), 83–87.

[22] Goryunov V.: Finite order invariants of framed knots in a solid torus and in Arnold's J^+-theory of plane curves, "Geometry and Physics," *Lect. Notes in Pure and Appl. Math.* 184 (1996), 549–556.

[23] Gramain A.: Sur le groupe fondamental de l'espace des noeuds, *Ann. Inst. Fourier* 27 (1977), 29–44.

[24] Goussarov M., Polyak M., Viro O.: Finite type invariants of classical and virtual knots, *Topology* 39 (2000), 1045–1068.

[25] Hatcher A.: Topological moduli spaces of knots, arXiv: math. GT/9909095.

[26] Hatcher A., McCullough D.: Finiteness of classifying spaces of relative diffeomorphism groups of 3-manifolds, *Geom. Topol.* 1 (1997), 91–109.

[27] Johannson K.: Homotopy equivalences of 3-manifolds with boundary, *Lect. Notes Math.* 761, Springer, Berlin (1979).

[28] Jones V.: Hecke algebra representations of braid groups and link polynomials, *Ann. Math.* 126 (1987), 335–388.

[29] Kashaev R.: The hyperbolic volume of knots from the quantum dilogarithm, *Lett. Math. Phys.* 39 (1997), 269–275.

[30] Kashaev R., Korepanov I., Sergeev S.: Functional tetrahedron equation, *Theor. Math. Phys.* 117 (1998), 1402–1413.

[31] Kauffman L.: *Knots and Physics* (Second edition), World Scientific, Singapore (1993).

[32] Kauffman L.: Virtual knot theory, *European J. Comb.* 20 (1999), 663–690.

[33] Kuperberg G.: Detecting knot invertibility, *J. Knot Theory Ramif* 5 (1996), 173–181.

[34] Mortier A.: Combinatorial cohomology of the space of long knots, *Alg. Geom. Top.* 15 (2015), 3435–3465, World Scientific, Singapore.

[35] Mortier A.: Finite-type 1-cocycles, *J. Knot Theory Ramif.* 24 (2015), 30.

[36] Mortier A.: A Kontsevich integral of order 1, arXiv: 1810.05747.

[37] Morton H.: Infinitely many fibered knots having the same Alexander polynomial, *Topology* 17 (1978), 101–104.

[38] Motegi K.: Knotting trivial knots and resulting knot types, *Pacific J. Math.* 161 (1993), 371–383.

[39] Murakami H., Murakami J.: The colored Jones polynomials and the simplicial volume of a knot, *Acta Math.* 186 (2001), 85–104.

[40] Polyak M., Viro O.: Gauss diagram formulas for Vassiliev invariants, *Internat. Math. Res. Notes* 11 (1994), 445–453.

[41] Przytycki J.: Skein modules of 3-manifolds, *Bull. Polish Acad. Sci. Math.* 39 (1991), 91–100.

[42] Sakai K.: An integral expression of the first non-trivial one-cocycle of the space of long knots in $\mathbb{R}^3$, *Pacific J. Math.* 250 (2011), 407–419.

[43] Thurston W.: The geometry and topology of three-manifolds, http://www.msri.org/publications/books/gt3m/.

[44] Turaev V.: The Conway and Kauffman modules of a solid torus, *J. Soviet. Math.* 52 (1990), 2799–2805.

[45] Turchin V.: Computation of the first non-trivial 1-cocycle in the space of long knots, (Russian) *Mat. Zametki* 80 (2006), No. 1, 105–114; translation in *Math. Notes* 80 (2006), no. 1–2, 101–108.

[46] Vassiliev V.: Cohomology of knot spaces. In: *Theory of Singularities and its Applications, Adv. Soviet. Math.* 1 (1990), 23–69.

[47] Vassiliev V.: Combinatorial formulas of cohomology of knot spaces, *Moscow Math. Journal* 1 (2001), 91–123.

[48] Waldhausen F.: On irreducible 3-manifolds which are sufficiently large, *Ann. of Math.* 87 (1968), 56–88.

[49] Willerton S.: *The Kontsevich Integral and Algebraic Structures on the Space of Diagrams*, Knots in Hellas 98, Series on Knots and Everything 24, World Scientific (2000), pp. 530–546.

[50] Willerton S.: On the Vassiliev invariants for knots and pure braids, PhD thesis, University of Edinburgh (1998).

Index

Printed in the United States
by Baker & Taylor Publisher Services